ANTICHRIST, COMPUTERS, AND YOU!

HIGH TECHNOLOGY AND THE BEAST

ALPHUS L. HARRIS

JOE H. E. SKATES PUBLISHING
SHELBY, NORTH CAROLINA

Published by:

    Joe H.E. Skates Publishing
    Post Office Box 848
    Shelby, North Carolina 28150

All rights reserved. No part of this book may be reproduced in any form or by any means, mechanical or electronic, including information storage and retrieval systems without express permission in writing from the author or publisher, except for short passages in a review.

Copyright © 1987 by Alphus L. Harris

Printed and manufactured in the United States of America.

Library of Congress Cataloging in Publication Data.
Harris, Alphus L., 1943-
    Antichrist, Computers, And You!
    : High Technology And The Beast
    Bibliography: p
    Includes index.
    1. Technology--Religious aspects.
    2. Technology--Social aspects.
    3. Antichrist.
    I. Title.
BL265.T4H3   1987   291.1'75   87-90642
ISBN 0-942989-49-X (pbk.)

TO
MY PARENTS,
LEONARD AND LUCILLE HARRIS,
AND TO
DORIS N. JONES

ACKNOWLEDGMENT

Many people have contributed to this book in many ways from just listening to me talk about it to reading and editing. In addition to Doris Jones and my parents, I truly thank Renée Allen Krishna Bappanad, Charles Bell, Mildred Bell Richard Bonacci, Bernie Cain, Freida Cochran Rob Dembiniski, Yvette Fox, Myra Goforth, Shannon Goforth, Wilford Goforth, Dorothy Hammond, Cindy Harris, Joann Harris, Maureen Harris, MindyHarris Wendell Harris, Nancy Johnson, Gordon Kromberg, Teresa Laughlin, JayMarcom, TexxMarrs, WandaMarrs Jim Mattoon, Lou Menendez, RuthMesser, Greg Moody Abdullah Nassim, Patrick O'Meara, Mike Press, Rhonda Press, JohnReynolds, Joan Murray Reynolds LindaRobinson, MableSarratt, JerrySutton, Phyllis Sutton, ChrisSutton, KevinSutton, LeBron Sutton Armen Tashdinian, Lillian Tobias, Betsy Tritley and Stanley Tritley.

And, a very special thanks to Reverend John Wesley White for permission to quote his very interesting and informative book, <u>The Coming World Dictator</u>.

NOTE

This book was written to inform the readers of some products and situations that exist or are evolving with technical significance, and seem to fit into the scenarios of some philosophies of the end times of planet Earth. It was not written to discredit nor to impugn anyone, any written work, any religious organization, nor business organization.

In many cases illustrations may not be exact due to the constant change in technology, politics, economics, or other areas of our own lives. Such change may include the passing of a President, war, or economic change as in recent airline mergers.

Fundamental, fundamentalism, fundamentally, charisma, charismatic, and many other words are used as they are basically defined and do not refer to a specific person, church, group, nor denomination.

Joe H. E. Skates Publishing and the author express that the only purpose of this book is enjoyment and to provoke thought.

TABLE OF CONTENTS

PREFACE		vii
THE BASIS		9
CHAPTER 1	THE SETUP	13
	Antichrist	13
	Kingdom of the Antichrist	21
	Scarlet Whore	24
	Beast	27
	The Mark	29
CHAPTER 2	AS THE BEAST ROAMS – SNEAKY, FLIRT, HIS GOAL, HIS WHORE	31
CHAPTER 3	YOUR HAIRS ARE NUMBERED	35
	The "BEAST"	36
	Bury The United States	40
	Your Hairs Are Numbered	42
CHAPTER 4	SIGN OF THE TIMES	45
	Cosmetics and Fads	56
	The Room of Horror	58
CHAPTER 5	FAULTLESS CONTROL	63
	Foolproof Identification	65
	Quality of Life	69
	Three or Four Digits	70
	Time Line	72
	You Support The Beast	75
	The Silicon Wall	77
CHAPTER 6	HANDSOME ANTICHRIST GIVES THE GOOD LIFE	81
CHAPTER 7	ANTICHRIST AND THE AIRLINES	89

CHAPTER 8	COMPUTER TALK EASIER	99
	Degeneration of Social Conditions	101
	Organizations, Governments, And The Beast?	102
CHAPTER 9	THE BEAUTIFUL, UBIQUITOUS BEAST, HE'S EVERYWHERE, HE'S EVERYWHERE	105
	Orwell, Bible Writers and the List	112
	Then Came Banks	115
	Universal, World Money	119
CHAPTER 10	WORLDWIDE SHOPPING	121
CHAPTER 11	WORLD CITIZENS ACCEPT THE ANTICHRIST	127
CHAPTER 12	ANTICHRIST SOCIAL CLIMATE	135
	Sex Survey	136
CHAPTER 13	PHONEMES, HOLOGRAMS, AND DIRTY POLITICS	143
	To The Nude Beach	146
CHAPTER 14	THE SPIRIT OF THE ANTICHRIST IS HERE	153
CHAPTER 15	BILLBOARD AND THE CHILDREN	171
REFERENCE AND READING LIST		177
INDEX		181

PREFACE

This book was written to describe the Antichrist, the Beast and the end of the age in which we live. It is written to cover three areas of interest:

    1. The first area of interest is a description of the fundamental interpretation of the Bible prophesies as I recall them being taught to me as I grew up.
    2. The second area of interest explores the world of modern day in the areas of politics, religion, war, economics, love, hate, and social trends as they relate to or tend to support the philosophy of the Antichrist rendered by some fundamental believers.
    3. The third area of interest is the mechanism or the technology to support the theory of the Antichrist as I percieve the fundamentals to believe it.

In general, my goal for this book is to provide worthwhile reading matter for:

    1. The person who wishes to know more about this philosophy of a world dictator and the end times.
    2. The person who wishes to further his understanding of computers and other high technologies and their places in our lives.
    3. The person who wishes to explore philosophy, sociology and anthropology in general.

    My ultimate goal for the book is to provide a forum of understanding of the fundamental Christian. My goal is to allow people to know that the fundamental Christian of my childhood, may have had different religious viewpoints, but was a thinking person. I would hope to dispell any idea that all fundamental or born again Christians are like some of those who were involved in the 1987 "HOLY WARS."

I have taken extreme care to keep my personal beliefs from entering into my discussion of the subject matter. In many cases statements seem to be very emphatic. These statements are not to be taken as an effort to prove to anyone that the philosophy is, or is not true. It is merely a rendition of my understanding of how people that I have met over the years felt about this subject.

My pleasure and a feeling of success would be derived if I knew that someone had read this book and was kinder to another person because of it.

ALH

INTRODUCTION
THE BASIS

Six, six, six!

Are you afraid?

Many are!

Maybe you should be!

Do you know why, or are you ignorant?

Six, six, six!

666 666 666 666 666 666 666 666 666!

The footsteps are here and those are the tracks -

    666!

The BEAST is on his way. He's getting closer every day.

Do you believe that?

When you finish this book you may! Or, you may not!

How do I know that 666 is the footprint of the Beast?

My Daddy told me so. He tells me now! He told me many years ago. He told me often when I was just a boy.
    Now I'll tell you the stories my father

told me. While telling you the stories that he told me, I will tell him and you the very intriguing parts of the stories that only the world of technology provides.

This book is an explanation of a fundamental Christian interpretation of the Antichrist and the end of the world as we know it today.

Years ago the whole philosophy as interpreted by fundamental Christians seemed ludicrous or the folly of the weak-minded. To many it was merely signs of the ignorance and narrow-minded view of the fundamental Christians. Today, high technology makes the Beast a feasible creature.
　　The religious philosophy alone is very interesting! The technological part is completely intriguing! Computer and communications technology scares many people, regardless of their positions in life. The following descriptions are meant to be helpful and not confusing.
　　If necessary to avoid confusion and misunderstanding I will repeat an explanation several times.

Intermingled with my father's stories are Bible references. Explanations of modern technology appear to give credence to the existence of the Beast of my father's stories.
　　They are interesting for the believer and the nonbeliever. When you complete this book you will know more about the Beast and the high technology world in which we live.

If you ask, "Are you a scientist?"
I will say, "No!"
　　If you ask, "Are you a preacher?"
I will say, "No!"
　　If you ask, "Do you believe what you are writing?"
　　I will not answer. I am not going to make any commission of my personal belief!
　　You answer the questions!

The Basis 11

Does it all make sense?

My major aim is to place a religious philosophy, which I was taught, beside a technology which I have experienced. The religious philosophy is one concerning a mighty beast. This mighty beast is not one that roams around in the jungle. The Beast is a mighty one that controls human lives through politics, economics, religion, and all other facets of human existence.

Twenty or thirty years ago it was easy for the critic or the 'doubting Thomas' to smirk at the idea of the Beast.

It was easy to say,"That's impossible."

Today, it is not so ludicrous. In fact, to those who were taught the theory of the Beast, it is extremely plausible. To many it is downright frightening. The whole idea and its practicality make many shudder. To many others it is still the folly of the ignorant.
    Just wait! When you have finished this book you will at least be less ignorant!

You may be aware of the religious philosophy of the Beast or you may be aware of the developments in technology. But, very few people know both the religious philosphy and the technology.
    There are many surprises!
    In addition to the interesting revelations in religious philosophy and high technology, the Beast also intrigues us with world topics such as anthropology, politics, finance and other century old evolutions which we sometimes prefer to forget.

To begin, it will help to clarify some important terms such as:

ANTICHRIST
KINGDOM OF ANTICHRIST
SCARLET WHORE
BEAST
MARK

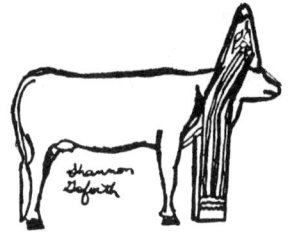

In any philosophy there are no absolutes. The following definitions are certainly not absolute. They will, however, give a basis for the general discussion.

CHAPTER 1
THE SETUP

ANTICHRIST

The Antichrist is thought to be a being that will rise up in world power and oppose Christ. A less popular thought is that the Antichrist is not a being, but merely a political force. From the viewpoint of some fundamental Christians any Moslem, Hindu, atheist, or believer of any other non-Christ philosophy would be antichrist. This viewpoint means that anyone who does not support the philosophy that Jesus Christ is God is antichrist.

In the 19t3 edition of Scribner's, The Dictionary of the Bible, Antichrist is defined as:
"The great opponent," or "counterpart" of the real Messiah, who finally conquers him.... "the idea of" one who persecutes the people of God, but, by the true Messiah, will be ultimately conquered .... (see Daniel 11:26 and Zechariah 12:14).

"...super-human figure," with the power and character of Satan, who was to head the opposition to Christ Himself and to the people of Christ. He was not to be "a general tendency," he was to be a "definite personality."

From The New Catholic Encyclopedia (1966) we have:
The Antichrist is going to be "an individual person...." He is one preserved for the final or "last times," his tyranny to "extend" until "the second coming" of Christ.

The term Antichrist many times produces more ominous thoughts than just an opposing or different belief. It fundamentally engenders thoughts of forced rejection of Christ. It even suggests punishment if Christ is not denounced.

Many fear the type of punishment described in chapter six of the Old Testament book of Daniel. In this story, the prophet was thrown into the lion's den because he continued to pray to God three times each day despite King Darius' decree forbidding anyone to pray to God. Everyone was to pray to King Darius.

Picture yourself!

The soldiers have you at the edge of the lion's den. The hungry lions are growling. Would you say, "Dear God...." Or, would you say, "Dear King Darius?"

This story is a good example of a high political official being duped into something he totally opposes. In this case King Darius liked Daniel. He had no desire to harm Daniel. However, at the suggestions of the jealous satraps, or governors, under him, he put out a decree. The decree had been devised by the satraps who were envious of the foreign prophet Daniel because he had found favor with their King.

Also, in the book of Daniel, the three Hebrew Children: Shadrach, Meshach, and Abed-nego, were thrown into a fiery furnace for failing to obey the King's decree. The decree ordered them to pray to a golden image instead of their omnipotent God, Yahweh.

Do you remember the song, "They wouldn't bow?"

Would you? Would your friends?

Like King Darius, the Antichrist will forbid you to pray to God. He will require you to bow to him. There will be an outward sign to indicate that you support and worship the Antichrist. That sign is the Mark of the Beast which is described on page 29. As under King

The Setup 15

Darius, those who fail to follow the rules and refuse to accept the Mark will be punished. However, the Antichrist punishment will be different, but will be equally horrible. You will not be able to buy food or obtain medical treatment. Through the use of high technology the punishment will not even be given by a human being. Many people believe that the Mark of the Beast will be forced upon the people as was the decree of King Darius.

This is probably not true!
The Mark will be readily accepted for the benefits it provides.
The Antichrist will control everything. He will provide so many good things of life that few will reject anything that he asks.

How will the Antichrist obtain such power?

As any leader who has not inherited his power, the Antichrist has charisma! By his looks, his speech-making ability and other such characteristics he will be able to get people to follow him regardless of his true evil nature.

Evil nature?

Yes, the Antichrist has two personalities.

One personality is very good and will last for the first three and one half years of his reign. In the second half of his seven years he will show a very different personality. To obtain the support of the world he must keep his evil nature hidden until he has supreme power.
For the first three and one half years the Antichrist will appear awesome and will display auras of godliness such as that in recorded instances of God appearing to man.

In the Book Of Mormon (Copyright 1963, pub 1974) a description of its origin tells how

Joseph Smith on the night of September 21, 1823 sought the Lord in fervent prayer. As he prayed there was a divine manifestation, a messenger from God. That messenger, Moroni, had been sent by God to tell Joseph Smith of The Book Of Mormon written on gold plates. He was there to reveal to Smith his mission. Smith described the messenger:

As wearing a robe exceedingly white, and his whole person being glorious beyond description, and with a countenance like lightning. Even the exceedingly light room was not as bright as immediately around his person. Smith was at first afraid; but the fear soon left.

In the book of Daniel in the Bible, Nebuchadnezzar describes the man in the fiery furnace with the three Hebrew Children as appearing like the Son of God.

In chapter 22 of Revelation, John writes that when the Lord arrives after the Great Tribulation there will be no need for candles for the Lord himself gives off the radiance and brightness of light.

Almost always, a manifestation of God has been described as a brightness that illuminates the surroundings. The event strikes the human observer with fear and then a settled peace.

So will be the traits of the Antichrist!

The Antichrist will be very, very handsome. Clark Gable, Rock Hudson, Rudolf Valentino, Robert Redford, Mel Gibson, Richard Chamberlain and others of handsome fame will not compare to the attractiveness of the Antichrist. He will be suave, eloquent, erudite, passionate, and charismatic. According to his followers only good adjectives can be used to describe him. He is the epitome of godliness!

He must be to fool the Jewish people.

The Setup

In chapter 13 of the book of Revelation we read that he goes about doing great wonders. He will "...bring fire down from Heaven." Further, the Antichrist will be able to perform miracles. Many fail to realize that the Bible does give the Devil credit for having powers. Scripturally, the Occult is real. Satan performs miracles through his servants.

The Antichrist will be or pretend to be a descendant of one of the twelve tribes of Israel, probably of the tribe of Judah. He will have charisma and solve world problems. He will be compassionate. He will provide a peace for Israel. The Antichrist will be an example of intelligence, fortitude, and strength. He will have miraculously recovered from a death dealing blow to which others would have succumbed (Revelation 13:3,12). To his followers he is the Messiah!

For three and one half years the Antichrist will be everything good and right. He will appear to provide everything for which you could hope. Times will be so good that people will relinquish the freedoms that democratic governments have so long tried to provide. Why does anyone need freedom if the dictator is so benevolent? Except for the "ready" Christians there is no reason!

But, for those not willing to accept the Mark it will be bad! The Antichrist personality of the last three and one half years is the personality most often thought of when one thinks of the Antichrist. It is the darker side of his personality. Ivan the Terrible was good by comparison. The Bible is replete with descriptions of this personality such as:

Son of Perdition – 2 Thess. 2:3,4
Mad man of Sin – 2 Thess. 2:3,4
The Prince of the power of the
    Air – Eph. 2:2
The spirit that now worketh in
    the children of disobedience
    – Eph. 2:2
Imbued with Satanic power
    – Rev 13:3–5
Blasphemous – Rev. 13:5

With the good and bad descriptions of the Antichrist, many fundamental believers over the years have asked, "Who?" Could it be: John Kennedy, Golda Meir, Charles Manson, Sun Myung Moon, or one of many others? One person recently even asked if I thought I could be the Antichrist. People have tried hard to make some important person into the Antichrist.

In the mid fifties even Dwight D. Eisenhower was mentioned. He was well liked and renowned, and some claimed that he was of Jewish heritage. Although he may have possessed many characteristics of the Antichrist, I found no support for his Jewish heritage in my research. Although Jewish heritage or the pretense of such seems necessary for the Antichrist, people seemed to forget that fact after Eisenhower.

John F. Kennedy of Irish Catholic descent was certainly not Jewish and did not pretend to be Jewish. Yet, in the early sixties Kennedy was often discussed as a possible candidate for the Antichrist. Many people brought up the name of President John F. Kennedy because he was so charismatic. Another reason was the fundamental belief that the Roman Catholic Church will play a major role in these end times. Therefore, many felt that Kennedy, the first Roman Catholic President of the United States, was the likely representative of the Roman Catholic Church. These facts for the fundamental believer were reasonable, but there were other things about

The Setup 19

the young President that did portend the world dictator more than just being Roman Catholic and charismatic.
   Some of President Kennedy's activities in carrying out his duties as President were somewhat disturbing. But, disturbing as they were, his activities seemed to go unheeded by the same fundamental groups that are so dogmatic in their espousal of the coming Antichrist. It appears that on February 16 and 27, 1962 President John F. Kennedy was provided complete and dictatorial authority to undertake immediate and decisive action in the event of a full confrontation.
   This authority was stated in a group of Executive Orders (EO) formulated during the Cuban Missile Crisis. These Orders were to be carried out through the Office of Emergency Planning and were to be put into effect in time of increased international tension or economic or financial crisis.

These orders seemed all inclusive. Basically, they were:

- to take over all communication media. EO 10995

- to take over all electric power, petroleum, gas, fuels, and minerals. EO 10997

- to take over all food resources and farms. EO 10998

- to take over all methods of transportation, highways, and seaports. EO 10999

- to mobilize civilians and work forces under governmental supervision. EO 11000

- to take over all health, welfare,and educational functions. EO 11001

- to operate a nationwide registration of all persons through the Postmaster-General, a member of the President's Cabinet. EO 11002

- to take over all airports and aircraft. EO 11003

- to take over housing and finance authorities, to relocate communities, to build new public funded housing, to designate areas to be abandoned as unsafe, to establish new locations for populations. EO 11004

- to take over railroads, inland waterways, and public storage facilities. EO 11005

- to designate responsibilities of Office of Emergency Planning, give authorization to put all other executive orders in effect in times of increased international tension or economic or financial crises. EO 11006

Although the powers of the President through these Orders are certainly the types of powers that the Antichrist will eventually possess, we like to think that our senators and congressmen will not allow us to fall into such a precarious situation, but it happened with these orders in 1962 during the national panic of the Cuban Missile Crisis.

Fortunately, some of our congressmen do see the dangers in such blanket authority for any one president. In 1969 through 1978 the Executive Orders listed above were revoked by Orders 11556, 12046, and 11490.

Recently, a friend of mine in a high government position said that we are farther away from accepting such dictatorial control than we have ever been. But, are we? People demonstrate and wave banners to indicate that they will not follow a leader while in many

The Setup 21

instances the demonstrations themselves are merely instances of the people blindly following a hero, public figure, or organized group. Do the Executive Orders above indicate that we will shun control?

In a "national hype" situation it appears that people will follow anything. People will surely follow someone as charismatic and handsome as the Antichrist. It seems an easy task for one so great to entice people to worship him. It will take little effort to get people to adore such a phenomenal being. All that he will ask you to do to prove your adoration is to accept his Mark, his adoration code - 666. You may even do it with an invisible tatoo, but many will want everyone to know that they are a member of the "in" society. They will have the adoration code in bold print across their foreheads.

    Bow down before him,
    Stand up,
    Adore him.

This will be the theme!

KINGDOM of the ANTICHRIST

The Antichrist must form a government. "Where will his seat of government be?"

Revelation 17:12 describes the kingdom of the Antichrist as the combination of ten kingdoms. Some believe that it will be ten federated nations of the European Economic Community (EEC)! Possibly the actual seat of this kingdom is Rome, The City of Seven Hills, which fits the geographic setting of seven mountains in Revelation 17:9.

Yes, the EEC, also known as the European Common Market or the World Common Market, is believed by many to be a major part to the Antichrist puzzle.

Why the European Common Market?

With a few more changes it will be a perfect fit for many old time prophesies. The gestation or the birth and development of the Beast and the seat for the Antichrist seem to parallel the evolution of the EEC. The forming of the government seat for the Antichrist is to be the coming together of ten nations to form one government body. This formation is an effort to control world politics, religion, and economics. The Antichrist government is also expected to take in the old Roman Empire. Some areas which are believed to be part of that Empire are: Italy, Spain, France, England, Holland, Turkey, Belgium, parts of: Hungary and Austria, many middle east countries on the Mediterranean Sea and some of North Africa. (reference: Larkin, Clarence, <u>Dispensational Truth</u>, Rev. Clarence Larkin Est., Phila, Pa.)

The European Economic Community was begun in 1958 with France, West Germany, Italy, Belgium, the Netherlands, and Luxembourg. These six nations were joined in 1973 by Britain, Denmark and Ireland. In the years since 1973 there have been frequent news articles that make the EEC seem closer to the Early Roman Empire and to the prophetical description of the Antichrist seat of government.

The charter of the EEC calls for:

1) the abolishment of import quotas and tariffs on goods traded among member nations -that is, a free trade area;

2) the creation of a system of tariffs applicable and common to all goods of non-member nations - eventually a trade barrier;

3) the free movement of labor and capital within the Common Market, and

The Setup   23

   4) the implementation of policies with
   respect to other areas of mutual interest
   such as agriculture, transportation, and
   restrictive business practices which are of
   concern to all member nations.

   (See Reading And Reference List: McConnell.
   Spencer. Samuelson. EEC Figures.)

The EEC, often referred to as the TEN, The Ten
of Europe, and Europe of Ten, does seem like
the government of Revelation.
    The EEC, with a population approaching 300
million and a combined Gross National Product
near $3 trillion is a powerful economic and
political force. It has accomplished much of
the goal set forth in the charter.
    Knowing that the Antichrist is suave and
handsome and that he reigns in Rome, you may
ask, "What will he do?"
    Using an old cliche, "When in Rome, do as
the Romans do!"
    Almost always when we think of Rome we
think of romance. The Antichrist is no
different. He will have a type of political
romance. Ultimately the romance increases his
power. But, wait! A real romance must have two
parties. So does this one! The lover is a
whore, the Scarlet Whore!

SCARLET WHORE

The Scarlet Whore is believed to be a religious philosophy or a church body. Based upon Revelation 17:18, the most popular belief seems to be that she is a church body. Further defined, many fundamentals in the past have believed that the church body is the Roman Catholic Church. However, in recent years that attitude may be changing. But, the changes of ideas toward the Roman Catholic Church do not seem to have affected the belief and support for the original ideas of the powerful church organization.

Recently others have mentioned the possibility of the powerful church body being the Moonies or the Hari Krishnas. Not long ago I became aware of a belief that the Mormon church is a primary candidate. Probably there are others which will come to mind when the subject is mentioned. In any case this powerful religious body is expected to swoon people into the fold by deception. This deception, however, is not flagrant. In fact the deception will occur through the appearance of goodwill for mankind. Anyone denouncing such a "good" church or organization will appear to be antichrist himself.

Could the Roman Catholic Church really fit the description?

Many still think so! But, some think that one of the Eastern religions springing up in recent years is a better fit than the Roman Catholic Church.

Certainly, no name is called in the Bible prophecies, but many of the descriptions can be compared to the Roman Church. In recent years, even the benevolent leadership of Pope John Paul II may foment that comparison.

The Setup

You may ask, "Who else fosters such ideas

Could it be our President?

"Oh, No!"

Just maybe, "Oh, Yes!"

When President Reagan established diplomatic relations with the Vatican, he created quite a forum for discussion. Many religious leaders and politicians expressed a grave concern that this act directly opposed the United States Constitution provision of separation of Church and State.
    Beyond the Constitutional question there were many comments as to the credence and power that this act bestows upon the Roman Catholic Church. As prominent as the Roman Catholic Church seems in the discussions of the Antichrist, there still seems to be some doubt. One book recently published appears to proclaim a subversive force or network that may be the demise of true Christianity. That network is known as the New Age Movement. Could the network be the right setting for the Antichrist?

During the rise to power of the Antichrist there will be a beast of burden. The burden on the beast will be the Scarlet Whore. The Scarlet Whore is basically like any other whore. She has a service to render to the people. The service in this case is a religious philosophy or a religious support system. The analogy or comparison to the whore is drawn due to the fact that the Scarlet Whore has the ability to provide a virtuous service to the world, but has succumbed to mundane or worldly pressures. She has bowed to her own vanity and desire for POWER!
    The desire for power has been the driving force for many persons throughout history. In chapter four of the New Testament book of Luke, Lucifer tempted Jesus to worship him in

exchange for earthly power. From then through modern times, leaders in almost every country have committed atrocious acts to attain power. Yet it is not just government officials. Take a look at religious groups that are in no way affiliated with the Roman Catholic Church.

What about Jerry Falwell, Terri Cole-Whitaker, Billy Graham, Jim Bakker or scores of others that are often in the news? Whether any one of these truly seeks power, it appears that way at times.

The Scarlet Whore, or religious body, will have gained vast power by the worldwide membership she enjoys. Daily news articles yield much information to support the idea that the Roman Catholic Church is the religious body. A very good example of this support is the abundance of the Pope's humane deeds. He visited Agca, the man who alledgedly shot him. Lovingly, the Pope forgave his assailant as a brother, displaying the true Christian spirit. He addresses the needs of the hungry. He attended mass in the Lutheran Church. He gives an audience to national ambassadors to the Vatican. He addresses world security. He, in fact, does seem to have the interest of the world at heart. To the masses this translates to a godly splendor. However, does this answer the question of the Roman Catholic Church or an Eastern religion?

No! But, a possible answer, was hot off the press in early 1986. Bookstore clerks told me that they could not keep the book titled, THE HIDDEN DANGERS OF THE RAINBOW by Constance Cumbey. This book seems to me to say, "Yes," to the question of Eastern religions. It seems to indicate that the Eastern religions may at least form a part of the body that subverts Christianity. Regardless of which church body it will be, 'The Church,' will have POWER and SPLENDOR.

Analogous to the ultimate Whore, 'The Church' uses this splendor to obtain what she wants which is to control the power of the

Beast. To obtain this power she bows to the wishes of the Beast. The Beast makes it appear to the Whore that she is getting a free ride to success. This frivolous appearance of success swoons the Whore and she mounts the Beast. The mount provides for two things. First it pledges the Whore to the Beast. Second, it unleashes the Beast.

Unleashed, the Beast runs rampant!

BEAST

So, what is this rampant Beast?

Revelation 17:3 defines the Beast as the mount for the Whore arrayed in purple and scarlet. Zondervan's <u>The Nave's Compact Topical Bible</u> defines the Beast as ...an "Apocalyptic symbol" of a "brute force," lawless and sensual, opposing God.

Symbolically, the Beast may be looked upon as the mascot of the Antichrist.
 Probably based upon Daniel 7 and Revelation 13:11-18, the Beast is thought to be a political power.
 Many think it is Russia. This idea stems from the location of Russia as well as the political disposition of the Russian government. Geographically, Russia occupies the Biblical locations of Gog and Magog which symbolically are forces of evil (Revelation 20:7-9; Ezekiel 38). Many others think that the political power may not be Russia, but another communistic government.
 In the Old Testament book of Ezekiel, Chapter 38 starting at verse 14, it appears to be a combination of governments which includes Russia. Here the prophet, Ezekiel, has seen the great nation of Gog from the North. With the armies of Gog are the armies of many nations. The number is so great that it is as

a cloud or shadow over the Earth. The prophet Ezekiel sees this massive army headed south to attack Israel, God's favored nation. As always, God will not let this aggression of Lucifer upon his people succeed even though he may allow them to be punished for their sins.

It is difficult to separate the Beast and the Antichrist. It may be more easily understood if the Beast is considered the political force and the Antichrist as the head of that political force or government. The Beast also maintains the mechanism of control which is the computer and other high technology. With this control, the Beast – Antichrist duo will rule the ten federated nations of the old Roman Empire which rules the world.

With the Beast, the Antichrist, and the nation defined, there have been many speculations as to how the governing body would develop. Many organizations or governments in the past were thought to be the beginning of that government. In each case many felt that the organization was ready to give power to the Beast as described in Revelation 17:13. Some of the organizations were:

        The United Nations
        The NATO
        The League of Nations
        The Trilateral Commission

It is easy to understand how these were considered. The first three in essence were the forerunners of the European Economic Community.

For the believer, the puzzle is almost complete! He knows of the Antichrist, his kingdom, the Scarlet Whore, and the Beast. But, what about the MARK OF THE BEAST?

The MARK

666 – a major facet of the Beast and the Antichrist.

For complete control, any ruler must have the support of his people. The MARK or Mark of the Beast symbolizes that support. Six hundred, three score and six! That simple number of man is the number of the Beast (Revelation 13:18). It is the adoration number, a symbol of belonging. It will identify the bearer as a member of the Beast clan!

So! Big deal!

What is the implication of the Mark? To what is the bearer entitled?
  The bearer of the Mark is entitled to food, medicine, recreation, cars, clothing, jewels, gold, silver, and eternal damnation.
  Yes! All of the above!
  The best of everthing until the last.
  During the Antichrist reign, only those with the Mark will be able to buy or trade anything! The Mark will be implanted in the hand or in the forehead (Revelation 13:16). If you don't have the Mark you will not be allowed to transact business of any kind. Transactions from checking into a hospital to buying stock will be refused. Refuse the Mark and you will see your children starve! You may immediately conjure up thoughts of some ghastly or BEASTLY creature that is ugly and mean. This will not be the case. But, rest assured, you that refuse the Mark will be refused medical treatment, food, clothing, and all that is necessary for life (Revelation 13:17).

CHAPTER 2
AS THE BEAST ROAMS
SNEAKY, FLIRT
HIS GOAL, HIS WHORE

During the beginning of the Antichrist reign there will be a courtship of the Beast and the Whore. The Beast will display all of his coquettish abilities. He will swoon the Whore and convince her that he believes her philosophy. He will agree that her offering is good for the masses, as she upholds the truth like no other church or social body ever has. To her delight, he will help her to overtake or amalgamate with other churches. Many believe this will be accomplished by an organization such as the National Council of Churches. To many this is a world peace organization while to others it is the demise of the inalienable right to worship as you choose.

In any case the Whore will become very powerful. She will be exactly what the Beast needs. Eventually he will take the Whore for a ride. In her state of infatuation she will mount the Beast.

She will attempt to gain the acceptance of the Beast. To do this, she will prostitute herself in order to gain the benefits possible only by becoming the bedfellow of the Beast. The Whore perceives these benefits to be the things which will eventually make her the World Church.

This union will be a mutual acceptance by the Beast and the Whore.

The Beast receives extra power through the endorsement of the Whore. By this time, the masses have had their religious beliefs so diluted and infiltrated by the occult, atheism, agnosticism and general indifference that religious allegiance is to anything that is "big and beautiful" and offers security and comfort.

The Whore receives power through the support of the government, the Beast.

Although there are mutual gains by the Beast and the Whore, we all know from the poet Homer near 900 B.C., from the Bible (Matthew 6:24, Luke 16:13) and from our personal experiences that there can be no two masters. Eventually, one becomes the boss. In mergers of big corporations there is one surviving corporation, usually meaning that one loses her power to the other. So it will be with the Whore and the Beast.

Almost as soon as the wedding takes place, the honeymoon will be over. Recent United States elections have exemplified such a short lived arrangement. When Ronald Reagan was running for President of the United States in 1980, it appeared that he tried to woo the religious right. Some of his campaign rhetoric made him sound almost as if he were a fundamental evangelist. Through his speeches and his criticism of the other candidates he was able to gain the support of the religious "Right" over Jimmy Carter, the Born Again President. Soon after the election some of his ardent supporters began to feel that he had used them.

It is quite interesting to see how easy it is to sway the masses. That same fundamental group seemed very upset with President Reagan in late 1981 and on to late 1983 because he seemed to have forgotten the religious right which had supported him. However, by early 1984 the same people were listening to him and proclaiming him their candidate for a return to the good ways of living.

A similar situation will exist with the Beast and the Whore. Once the Beast has swooned the Whore, he begins to act more like a pimp trying to bring his prostitute under control. The Whore may moan and groan to the Beast, but will comply to his every wish.

AS THE BEAST ROAMS
SNEAKY, FLIRT, HIS GOAL, HIS WHORE

You may ask, "Why will such a power as the religious Whore bow to the Beast?"
    The answer is simple!

Over the years the mighty power of the religious world has slipped. For many years the Whore (or church) has at times dominated the known world. Year by year that power has slipped. The ignorance spread and perpetrated by the Whore has fallen by the wayside for intelligent and educated people. At times her power wanes. At times her power is boosted by a strong politician who professes to believe her dogmas. Many people follow the charismatic leader and accept his beliefs as their own. This blind following of the leader has given the Whore periodic boosts in strength.
    The Whore, realizing her waning power, sees the Beast as a means of consolidating all her powers.
    Once the Whore has bowed to the wishes of the Beast, she is bound. She is controlled by the Beast through her fear of the credibility gap of her subjects or members. The credibility gap would be created by telling her subjects that she, the mouthpiece of God, has been outsmarted by the devil himself. She has been the herald of godliness and now to admit her prostitution activities with the Beast would destroy her. She is trapped!
    The Lady has lost her maiden head and has become the epitome of a whore, The Scarlet Whore.
    With the noose around the neck of the Whore, the Beast is free to roam.
    What does this mean? Where will he roam? Why will he roam?

   The Beast will roam everywhere --- to conquer!

The Beast will roam into your life, my life, and the life of everyone. He will roam freely carrying his mount, the Scarlet Whore.

Together they will carry out the will of the Antichrist. Although the Beast has conquered the Whore and roams freely, the Beast too is bridled. He is bridled by the wishes of the Antichrist. Together the Whore and Beast provide the needed public relations team for the Antichrist. Through this union, all that is necessary for the Antichrist to take his seat has been formed. This seems ridiculous. Many argue that no organization or body will ever gain such control. However, one only needs to think back to the Executive Orders which gave President Kennedy virtually unlimited power. Even if we believe that absolute power can be obtained by a world leader, we may still wonder:

"How will it happen?"

CHAPTER 3
YOUR HAIRS ARE NUMBERED

As we discussed governments, leaders, and nations we saw how they fit into the Antichrist puzzle. Yet, there is a significant part of the puzzle that is often missed. That is a group of money-mongering preachers.

The Antichrist puzzle has been the subject of many sermons. It has been preached vehemently, by many fundamental believers, for two reasons. The basic reason is the fear of God. The second reason, based upon that fear, is a strong belief that the "end-times" are very near. The sincere belief is that the truths of God have been deciphered. This accepted truth fuels the spiritual fire and the preaching continues from fear of God and the sincere belief that this is 'THE END.'

Many times, however, the philosophy of the end times is preached to keep the faithful returning with dollars, dollars, dollars. Often these money mongering preachers use scare tactics to keep their listeners listening and sending money.

One article, "The "BEAST"," seems to be an example of this type approach. This article was passed around the country from coast to coast. In one organization it was distributed to the individual church pastors by the head of the state organization. It reached me bearing no credit as to the author. For this reason it is printed below without credits to the author. Read it. Do you think the author was using scare tactics?

The "BEAST"

Dr. Randrick Eldeman, Chief Analyst of the Common Market Confederacy, announced today from Brussels that a computerized restoration plan is already underway in the aftermath of world chaos. In the crisis meeting which brought together scientists, advisors, and C.M.C. leaders, Dr. Eldeman unveiled the "BEAST". The Beast is a gigantic computer that takes up three floors at the Administration Building of Market Headquarters. This "MONSTER" is a self-programming unit that has over one hundred sensing input sources. Computer experts have been working on a plan to computerize all world trade. This master plan involves a digital numbering system for every human on earth. The number to use for all buying and selling to avoid the problems of ordinary credit cards. The number would provide a walking credit card system. The number would be invisibly "laser-tatooed" on the forehead or back of the hand. It would provide a walking credit card system. The number would show up show up infrared scanners to be placed at all check-out counters and places to business. Dr. Eldeman suggested that by using three six digital units, the entire world could be assigned a working credit card number. Other Common Market Officials believe that the present chaos and disorder caused by the "Mystery" points to the need of a world currency - perhaps an international mark that would do away with all currency and coin. Instead, credit notes would be exchanged through a world bank clearing center. No member could buy or sell without having assignment of a digital mark. Market directors are now convinced that world order depends on allegiance to an international program of peace and politics, as well as a new world trade and numbering system. One man could have at his finger tips the most powerful lever known to mankind. He could have solution for world problems. It could be a tool for peace, or a dictator weapon.

When one of the Market Leaders was asked what would happen if any person objected to the system and refused to cooperate, he replied rather pointedly, "We would have to use force to make him conform to requirements."

When we remember what the Bible says (Revelation 13:11-17) concerning the appearance of the Anti-Christ as the head of this organization, the following quotation becomes exceedingly illuminating. Henri Spaak, early planner of the European Common Market, and Secretary-General of NATO, said in one of his speeches: "We do not want another committee; we have too many already. What we want is a MAN, of sufficient stature to hold the allegiance of all people, and to lift us out of the economic morass into which we are sinking. Send us such a man, and be he god or devil, we will receive him." (Quoted from The MOODY MAGAZINE, March, 1974)

In the light of happenings such as this we should by all means "watch and pray that we may be accounted worthy to escape all these things that shall come to pass and stand before the Son of man." Luke 21:36

End of Article

****

I have tried to ensure that nothing, including grammar and spelling errors, was changed in the article. If the article quotes these world leaders accurately, it is definitely unnerving. They have the power to make decisions that will affect the whole world, but they lack the knowledge of what is happening in the world of technology. In many cases the people with the knowledge of the technology advocate giving our technical prowess to a group such as this.

Even Albert Einstein suggested that the secret of the atom bomb should be committed to a world government and that the United States should immediately announce its readiness to give it to a world government (reference: <u>The Day The Dollar Dies,</u> Cantelon, William, Logos International, N.J., p 132) This was in 1945 and was a response to a statement, "Either we will find a way to establish world government, or we will perish in a war of the atom." The idea of one world government persists. In the 1970's and the 1980's organizations such as the European Economic Community seem to be headed that way. Many think that is the goal of the Trilateral Commission.

From the standpoint of technology, if 'computer experts' are spending much time developing a digital system just large enough to give an identification number to each of the inhabitants of the earth past, present, and future, the title he gives them, 'computer experts', should be changed to 'computer idiots.' A digital system has been in existence for hundreds of years. It is the use of zero and the Arabic numerals one through nine. This system could be called Genetic Modelling if you wanted to impress someone with new terms and to make your religious postulations more mysterious to the reader. All this really means is that through genetic transfers most people born over the years were born with ten digits (eight fingers and two thumbs).

Voila!

A new born babe already has that digital system over which these 'computer experts' labor. The maximum number expressed by the three six digital units mentioned in "The "BEAST"" would be:

999999 999999 999999

Your Hairs Are Numbered 39

or, 1,000,000,000,000,000,000 - 1.

This number is approximately equal to two hundred million times the current world population of approximately 5 billion people. If you believe the Bible prophecies, you will probably admit that prophetically there is not enough time left for the world population to explode to use up all the identification numbers. If so, it would be easy to convert to the hexadecimal representation system. The hexadecimal system appends the first six letters of the alphabet A,B,C,D,E,F to the zero through nine. This yields 16 digits 0, 1, 2, 3, 4, 5, 6, 7, 8, 9, A, B, C, D, E, and F. If the human had seven fingers and one thumb on each hand this is the method of counting we would be using today. A comparison of the decimal system that we use everyday and the hexadecimal system just described is shown below:

DECIMAL	HEXADECIMAL
0	0
1	1
2	2
3	3
4	4
5	5
6	6
7	7
8	8
9	9
10	A
11	B
12	C
13	D
14	E
15	F
16	10

100 in hexadecimal notation equals 256 in our decimal system. 1000 hexadecimal = 4096 decimal. Do you see how numbers are expanded using hexadecimal? Understanding numbering

system concepts such as hexadecimal makes the people quoted in "The "Beast"" seem very naive.

"The "BEAST"" is like many sermons. Some fact, ignorance, questionable information, and an abundance of sensationalism!

In earlier discussion of the prophesies we saw that there will be ten nations coming together as one power or one nation. Some believe that power will be the restructured Roman Empire. As explained by Hal Lindsay in his book <u>A NEW WORLD COMING</u> there are several empires that have formed a chain of rule throughout world history. Most of the empires in this chain have risen and have been defeated, but there is one empire that remains undefeated although it declined and went out of existence. That empire WAS AND IS the Roman Empire. Through its own internal decadence, the Roman Empire faded from power without defeat by another power. As the European Economic Community develops, many see it as a reformation of the Roman Empire. Others strongly oppose that theory because of questions about the part currently existing powers such as the United States of America, Russia, China and England would play in the rule of the Antichrist.

John Wesley White in his book <u>THE COMING WORLD DICTATOR</u> seems to indicate that that is not a problem. The United States may play no part of importance within the reign of the Antichrist.

<u>BURY THE UNITED STATES</u>

In the nineteen sixties one person spoke of burying the United States and upset the whole country. Yes, Nikita Kruschev alarmed the United States by saying that he (implying Russia) was going to bury the United States. People assumed that he meant through war, but he later explained that he meant economically.

Do you think that anyone can bury the United States?

Absolutely not!

Only the United States can do it? Just as the Roman Empire did it to itself, the United States can do it to itself!

In the movie "TORA TORA TORA" a Japanese leader of the World War II era opposed the Pearl Harbor attack because he felt that it would merely wake a sleeping giant.
It was true then and it is true now, the United States is a sleeping or drugged giant!
Could any adversary conquer the United States today? It is doubtful; however, as the ancient Roman Empire, the United States will crumble from its own decadence and I am implying nothing sexual as many preachers of sensationalism. In fact it sometimes appears to me that many preachers are the major part of the demise and decadence of the United States and the denigration of Jesus Christ.

The conglomerate Antichrist nation will wield great power over the world economy. As mentioned in the discussion of the European Economic Community (EEC), a powerful government has been developing for many years.
Although there is some uncertainty as to how the Beast, the Scarlet Whore, the Antichrist, the EEC and other entities will join hands to obtain the proposed final results, there is another important question fundamental to the final results. That is, how will the Antichrist maintain the information necessary for this universal power?

You probably already know the answer to the mechanical part of this question. Fundamental or not it is on the tip of the tongue. You can readily answer,

"The computer!"

YOUR HAIRS ARE NUMBERED

The important question is: "Who or what has the right to this kind of information?" We would all like to think that the answer to this is, "No One." Yet, we know that the information banks are being created and are growing daily. The United States Federal code of law prohibits its intelligence organizations from collecting information on individuals that is not pertinent to the operation of that organization, but who defines pertinent. Forget the governments of the nations in existence today. Consider large corporations like Sears. What is the capacity of that conglomerate to maintain information. Do you have an account with any one of the Sears companies: real estate, stock brokerage; or do you just have one of their charge cards? Sears necessarily maintains information on individuals for business purposes; and, Sears is not the only corporation involved in such information processing and storage systems.

TRW also maintains a massive bank of information about individuals. TRW and its subsidiaries are involved in the manufacture of automobile parts, the manufacture of aircraft parts and many other areas of business including credit information.
 Many years ago the claim was that TRW was adding fifty thousand names to their Credit Data Bank each week and claimed that they would soon have a record of everyone that had ever used credit (reference: ibidem, Cantelon p 69).
 TRW and SEARS are only two organizations out of many. It is not just business and government. It is special interest organizations, political parties, churches, and even the Ku Klux Klan. Common knowledge among leaders of any organization is that the most important commodity in any endeavor is INFORMATION. The Information on each one of us continues to grow in the many data

(information) bases throughout the world. You may not think of yourself as important, but you are certainly important enough to be recorded somewhere, almost inevitably.

You are important enough to have "every hair of your head numbered." Do you remember how ridiculous that statement once seemed? In Matthew 10:30 Jesus said, "But the very hairs of your head are all numbered." Many thought it riduculous to take the statement literally. Sarcastically, many asked, "How would God ever keep up with such numbers?" First of all why would any God want to number the hairs on your head?

I cannot tell you why any God would want to keep such records unless his formula for making things only allows so much for hair. However, I can answer the other questions.

With the computer systems of today such records could be kept without needing the powers of God. The other question I can also answer. Again I cannot speak for God, but I can give you a reason that man is actually keeping records of the numbers of hairs on some heads. An article in the Washington Post on December 1, 1984, was about the results of a year long test of the drug minoxidil. The test was to determine if the drug helps a bald person to grow hair. The test showed positive results in that "... the drug treatment doubled the number of hairs in the target areas on 27 of the tested heads. No one, 'he (Dermatologist Thomas Nigra) said,' experienced any hair loss."

When these experiments began the hospitals could not accomodate the volunteers. The interest was mainly for vanity reasons; however, there are other reasons to keep records on the hair. One of those reasons is that deposits of certain minerals and chemicals such as arsenic in the hair can at times help determine the cause of death of a person. If we are not so myopic or short sighted we may see many reasons to maintain records on the hair. It is also interesting to note how such insignificant things as counting hairs can affect our lives. Almost always there is an economic impact!

In the case of the drug minoxidil the financial effect was almost phenomenal. Upon the announcement that the drug was successful in the test, the price of the stock of the company making the drug shot up. A few months later a second company announcing a similar hair growing drug also saw the price of its stock shoot up.

Like hair counting, many of our old, anthropomorphic (to place man like characteristics upon other beings especially upon God) ideas dissolve as we just sit and watch history in the making. Day by day as our questions are answered we can see that ONE WORLD GOVERNMENT!

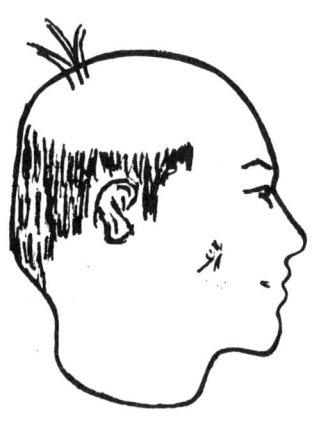

CHAPTER 4
SIGN OF THE TIMES

How will this one-being Antichrist-rule happen? How will the information gathering and the transfer of information from the Whore to the Beast be managed?

One perennial question seems to be, "How will you get people to take something as ominous and ugly as the Mark of the Beast?"

It will certainly take some public relations expertise. At least, that is how it seemed in the past, but the ugliness of the Beast is rapidly disappearing.

Another good question: "As efficient as government is, how will it maintain all of this information?" For years the answer to this question was elusive, however, the answers to both questions are here today. The public relations question will be answered later in discussions of sociology in the United States. The answer to the maintenance of the voluminous amount of information, as already stated, is, "THE COMPUTER!"

Yes, today at the hub of advances in technology, with its peripherals like disk packs, printers, CRTs, et cetera, the computer is the answer. The pieces of the puzzle are falling into place. Technological developments are so frequent and dramatic that they cause intrigue and fear. Biochips, genetic engineering and "Star Wars" make our magazines and newspapers read like science fiction. But, be assured, it is not fiction.

Descriptions of comments by world leaders such as those in "The "BEAST,"" assure us that those leaders are so ignorant of technology that they can never lead us into a world government; however, each day that goes by we see leaders becoming technically educated or supplanted by those who are literate in a technical sense.

In putting the Beastly puzzle together, the remaining parts of this book emphasize some of this and describe this "science fiction" in simple terms. The puzzle of the Antichrist involves history, politics, electronics, computers, communications and mass psychology.

The Antichrist philosophy is not to be brushed off lightly. When newspapers of repute from Los Angeles to New York publish articles concerning the issue, it is something to observe. Whether or not it was predicted by God, there is enough power in the masses for self fulfilling prophecy. The estimated group of 40 to 50 million United States residents who truly believe in the imminent appearance of the Antichrist seems to be growing. This helps to ripen the political, social, religious and technological climates for the Antichrist!

Long before I was born there were stories about the imminent arrival of the Antichrist. Those earlier stories of the definite and soon coming Antichrist were almost always based upon political events. Appropriately some of the stories began with a United States President!

That President was Franklin Delano Roosevelt, FDR!

FDR seemed to conjure up thoughts of devil, angel, and savior. When he first began his salvation programs for starving Americans, the ideas of the arrival of the Beast began to grow. The depression following the 1929 stock market crash opened the way for the arrival of the Antichrist.

In his blossoming years the Antichrist will show the likeness of Christ or the long awaited Messiah of the non-christians. He will clothe and feed the masses. He will put "A Chicken In Every Pot." Have you ever heard the famous FDR campaign promise?

Franklin Delano Roosevelt was a savior! Many people believe that he saved America; he was God's gift to mankind. His graces spanned the races, creeds and colors. Alas, Roosevelt was a paradox. One part of his nature seemed to contradict the other. He was a savior, but his political ilk frightened many because it fit the Antichrist description.

The Antichrist will be a savior! Immitating the real Messiah, he will provide prosperity, at least during the first half of his seven year reign; but, woe, only for those who accept the ultimately damnable MARK!

The word "accept" specifically implies that people will accept the Mark of the Beast without any visible means of force. Here, there is special meaning to the word force. Many will be forced to accept it, but there will be no demanding forces involved. Through technology and because of social conditions people will accept and even beg for the Mark!

Many believe that this subtle force or acceptance began with Franklin Delano Roosevelt.

Why FDR?

Do you remember your Social Security Number? Some believe that it is the MARK! But, immediately you think, "It is not 666." True, true, true! The Social Security Number has too many digits and uses all the digits from 0 to 9. Then, the Social Security Number could not be the MARK! In light of this fact, there are some concessions. Some people rationalized that 666 was not the actual number, but a symbol of a number such as the 'three six digital numbers' as stated in the article, "The "BEAST."" Anyway, concessions or rationalizations, the Social Security Number is not the Mark of the Beast. However, a "sign of the times," it has set the stage for the acceptance of the Mark.

Does this seem strange?

Yes, it seems as much a fantasy as space stories did just a few years ago. Flying around the universe with Buck Rogers was a whim or fancy of a child or a weak minded adult. Space encounters have become so real that 'Star Wars' became an issue in the Presidential election campaigns in 1984 and continue with us as we approach the 1988 elections.

Today most of us would laugh at anyone who would say, "No one has ever been to the moon." We would laugh upon hearing someone say that the earth is flat. It is unbelievable and could only be said by the ignorant or stupid. We forget that we are not all alike. We have not had the same experiences. Recently a young lawyer friend of mine told me that his Princeton University classmates were just like anyone else even though their parents are wealthy and well educated. We tend to forget that people differ drastically because we are most often around those that have backgrounds similar to our own. We lose touch with people in other stages of life. We tend to forget that as incredulous as it seems there are people who truly believe that no one has been to the moon and that the earth is flat.

One uneducated man I know argues that the earth is flat because the Bible speaks of four corners of the earth; if the earth is round it has no corners. He explains the rising and setting of the sun, "It rises and crosses the earth. In the evening it goes through a series of vibrations allowing it to cross back to its original Eastern position while remaining invisible to man."

Immediately the man is labeled ignorant. Yes, he is, but he is also ingenious. All of his intelligence is directed toward theories to support his fundamental belief. It is a very good example of how people justify or support their beliefs. It is the type argument that some dogmatic people use to try to impose their own beliefs upon the country as law. But, let's look at our own ignorance.

On the front page of the Washington Post, November 20,1984 there was an article about Genetic Engineering. The article was about some fantastic scientific experiments and developments which have produced very healthy laboratory mice by injecting an embryo with a human growth gene. With genetic engineering there are very positive implications in areas as divergent as cattle farming and the treatment of hemophilia or diabetes, yet there are those who oppose the research on the basis that we are crossing the species and giving human characteristics to animals.
 The scientist goes on to relate the fact that as reprehensible as it seems to the chronic complainers, wheat and rye which are not of the same genus are already crossed to produce a new grain called triticale. Near the end of the article biologist John Fulkerson is quoted, "Mother Nature has done more to commingle the genes of the species than we can ever do."

This kind of argument points out many things.
 First, if a person actually believes that the Bible is the inerrant word of God, it seems that he would follow some of the very simple teachings about prayer. EVERYONE can pray for corrections or inhibitions of the problems perpetrated upon mankind by the 'diabolic or devil possessed' scientist.
 Second, the fundamental Bible believer should also see these events as a 'sign of the times' and that his most important recourse is to stay in touch with God while spreading the 'Word.' The fundamental believer of the teachings of Jesus Christ has no time for political campaigns which are of so little eternal significance.
 Is the man who believes the world is flat the only ignorant one around? Some with

college degrees and even some who own or run colleges seem just as ignorant. It reminds us that only a few years ago television was the 'tool of the devil', anyone that was labeled a scientist was automatically the enemy of the 'God fearing society,' anyone that had an idea very different from the usual was considered wrong. Any thing or person that is foreign is to be hated or feared – surely a "sign of the times!"

Pulitzer Prize winner Peter Viereck who wrote about this in his paper "The Revolt Against the Elite" coined a word. That word is transtolerance. He said that transtolerance is a bridge between the words tolerance and intolerance. Transtolerance is the form that xenophobia takes when practiced by a 'xeno'." Xeno means foreigner. Phobia means fear. Therefore xenophobia means a fear of foreigners. Transtolerance is that fear of foreigners for other foreigners.

Earlier in this book we saw an example of transtolerance. Those criticizing the scientific advances of the day are afraid of the scientist who are seen as foreigners of the Godly forces. As in Viereck's statements, the only ones that have the fear of the foreigners are the other foreigners. A study of immigrating groups since this country began supports this idea. In very recent times Mariel Cubans were not readily accepted by Cubans who had arrived in the United States many years earlier. In the case of the people who are trying to impose upon others their beliefs which are based upon their own ignorance, they are just as foreign to the writings of God as the ones whom they criticize. In the same article Viereck states that America is getting closer and closer to racial equality yet is moving further from liberty of opinion. It seems that his statements become more true each day.

Although politics and science provide good bases for many arguments we do not have to go that far to see intolerance. We can look at

Sign Of The Times 51

the religious rituals for good examples. As the arguments of the man who believes that the earth is flat, many rituals followed daily in houses of worship are just as ridiculous to the outsider or the xeno.

Is it not ludicrous for a person to light a candle and pray to the Mother of Jesus as an intercessor to Jesus Christ. It seems so when the proposed Holy Writ of God speaks nowhere of needing an intercessor to reach God except the God-son, Jesus. Yet, to a xeno this ritual is carried to an extreme.

In Portugal at the shrine of Fatima, I watched the faithful walk around the shrine on their knees prior to entering to place a candle at the foot of a statue of the one honored as Holy.

Is it not beyond belief that a man on earth can say that part of the world should not eat meat on Friday; but, another section of the world may, through a dispensation, be exempt from the "meat eating" rules. It is especially strange when that dispensation rests solely on the actions of one man such as Simon Bolivar.

Is it not silly to believe that the Messiah is going to arrive someday and drink the glass of wine that is poured for him at each meal, especially on holidays?

Is it not ridiculous to think that a woman must depend upon her husband to get into heaven; therefore, the man may take more than one wife?

Is it not silly to think that a cow may really be an ancestor?

Yes, to someone not of the faith, whatever the faith, it is silly. Yet all of these "silly" practices and beliefs have formed integral parts of the rituals of religious groups in which we find world leaders, financial wizards, and scientists.

Often, when not of the faith, there is, to the other extreme, a shunning or disregard for other beliefs and rituals - another "sign of the times!"

Such disregard is seen in the actions of Winston Churchill. Condsidered a wise man, many respected his word and followed him into battle, even one over his desire to stop the circumcision of women in the Mau Mau community in 1955. This was one of the reasons for the uprising that claimed many lives. He thought he was dealing with heathens, but the heathen practice of circumcision of women has found its way into the Western medical world of the seventies and eighties.

Such political actions and religious beliefs lead us to the need for the WORLD GOVERNMENT even though the Beast may still seem to be a fairy tale. To Franklin Delano Roosevelt the thought that his programs, established to alleviate human agony, could be part of the work of the Beast would have been ridiculous. Yet, that as described on page 47 and 48 was what I heard as a child, not long after the Social Security System was implemented.

It still amazes me that my father told me those stories then and continues to tell them now. While telling those stories he had many questions. In the early days of the Social Security System there was the question of how the Mark of the Beast would be implemented. He felt surely that the Social Security Number was in some way the predecessor of the Mark.

He still believes that! He need not be uncertain!

He told me WHAT would happen.

Now, I can tell him HOW it will happen!

At least for three and one half years the Antichrist will not appear evil. The ones that will really seem evil and heartless are your friends and relatives. They will be like the friends of Job in the Bible. They will refuse to help you. They will even scorn you for not accepting the Mark. If they help you they will jeopardize themselves. They will simply admonish you and tell you that you could save yourself some suffering.

It's simple!

Take the Mark!

The part that may be hard to believe is that you will likely take the Mark if you are on the Earth at that time.

Yes, YOU will likely accept the Mark,

```
       666              666              666
    66    66          66    66         66    66
    666   666         666   666        666   666
    666               666              666
    666   666         666   666        666   666
    666 66666         666 66666        666 66666
    66666    66       66666   66       66666   66
     6666    66        6666   66        6666   66
      666    66         666   66         666   66
       66    66          66   66          66   66
          6666              6666             6666
```

---unless you are Jewish!

Yes! Allegedly some Jews will not accept the MARK. Some teach that many Jews will "see the light." Ezekiel 20:34-38, 22:19-22 seem to indicate that one third of the Jewish population will realize that when they rejected Jesus Christ two thousand years ago they rejected the true Messiah. Upon realization of the mistake they will accept the consequences, even death, to avoid the Mark and eternal damnation.
But why the Jews!
Many Christians believe the Jews are God's chosen people. Therefore God will protect them

and give them a chance to correct their mistakes. To these Christians, the Jews have made only one major mistake in the past two thousand years. That was the rejection of Jesus Christ as the Messiah.

This belief may not be understood by Jews. A Jewish friend from New Jersey once told me how Southerners hated the Jews. He hesitated to visit in my parents' home for this reason. It was quite a shock to me. At that time I had never known a Southerner who hated Jews. Those whom I knew well enough to know their sentiments toward Jews felt that Jews are God's chosen people. Southern or not, that thought seems to be somewhat prevalent among Christians. The idea certainly seems to be supported at the national political level. About fifty percent of the United States foreign aid goes for Israel. Some would say, "Yes, but that's because of the Jewish political influence in the United States government."

No doubt there is Jewish influence in politics, however, the Jewish population of the United States is only approximately three percent of the total United States population.

Regardless of the percentage of the population, the Jewish influence is greater than the numbers indicate. Look in your telephone directory. Count the number of Jewish doctors, lawyers, business owners and other prominent persons. The next time you watch television or a movie take note of the directors and producers. From numbers only the Jewish community could not accomplish so much. Yet, they have accomplished much. That accomplishment in part is due to the support of Christians. Without the support of the United States, Israel could not survive. Many will not even listen to negative statements about Israel in political situations. To many, Israel has the right to do anything because they are "God's Chosen People."

The fact that some Christians blindly

support anything Jewish in no way diminishes the accomplishments of the Jewish people. Read the <u>Diary of Anne Frank.</u> It dipicts only a small portion of the horrors which Jews have suffered throughout history and yet they have survived. Not only have they survived, they have prospered. They have prospered probably more than any other group of people. I feel sure that there are some ignorant, stupid and poor Jews, but I have never known one and more than three percent of my friends are Jewish. Thank God there are no government quotas on friendships.

Why so much talk about the Jews?

It is to emphasize their determination and will. It is also to emphasize the intellect and devotion of this small community of people. This drive and devotion at times may seem to others somewhat foolish. To gentiles some of the rituals of the Jewish religion may even seem silly.

Well, silly or not, they are symbolic of determination and perseverance. The traditional act of setting the glass of wine for the long awaited Messiah is an example of their devotion and patience in the wait for the Messiah. This long search and devotion may have kept the Jewish people aware of political and social situations around the world. It may be the reason they will recognize the Antichrist when he comes to power. It is quite interesting to note that all of the players in the cast built around the scenes of the Antichrist are Jewish. The Antichrist will appear Jewish, Jesus Christ was Jewish, those rejecting Jesus Christ as the prophesied Messiah were Jewish and the ones to recognize the Antichrist as an imposter will be Jewish. Many Christians believe God's timepiece is the actions of the Jewish people. The Antichrist is just another spot on that clock.

From this Jewish clock, it appears that

the whole picture is complete.

That is not the case!

Many seem to know exactly where the Jews fit, but cannot place the other pieces of the prophetical puzzle.
From the nineteen thirties through the seventies many simple pieces of the Antichrist puzzle remained unsolved. These remaining pieces were very important to the fundamental believer, but were often very basic. One of the basic questions was merely cosmetic. It was hard to understand how the Antichrist would convince a person to put a number in big black numerals across the forehead. Many thought it would just be another fad.

COSMETICS AND FADS

Would anyone follow such a fad?

Possibly! At times throughout history thousands have followed fads. In the Washington Post, Mary Hatwood Futrell, President of the National Education Association, in her discussion of computers in education stated that some consider the computer a technology - Hula Hoop, just another fad.

Just what is a fad?

My hand held Random House dictionary succinctly defines a fad as "a temporary, popular fashion, manner of conduct, et cetera."
More explicitly Futrell considers a fad "...something that fades without leaving any lasting imprint on the learning process."
By these definitions, the Mark of the Beast will not be fad unless one accepts the seven year reign of the Antichrist as a

Sign Of The Times

temporary period of time with no lasting imprint on history. Yet, it could begin as a fad. Some accept anything faddish even fads such as wearing a white glove on one hand as a popular entertainer has. Whatever the reasons, fads do occasionally sweep the country and the world. Thinking back about past fads, we are reminded how easily some people are led, yet, others do not follow so willingly.

Many times you may have worn one black sock and one brown sock because you were half asleep when you dressed. But, how many times have you worn only one white glove?

As the white glove fad, no fad reaches a majority of the population. It does, however, show a "blind following" trait of people. How many would follow the white glove fad if they knew its historical significance?

At one time it was punishment for young men in boarding schools. Boys were forced to wear one white glove when they were caught masturbating.

Probably, neither the young entertainer nor his followers know of this bit of history.

We tend to laugh off fads as quirks of the young and hope that adults are not so easily led. Yet, observing politics in this country we know that many adults form opinions based on very shallow reasoning. At times it is frightening to see what changes opinion polls about something as important as the presidential elections. Even beyond opinion polls and elections, we find evidence that people will follow a leader at any cost. People will follow the person perceived to have authority even if that perception of authority is based solely upon the dress of the individual.

Internationally there have been cries about the officers who served the Nazi regime under Adolf Hitler. Searches have continued since World War II to find the men who committed horrible atrocities against the Jewish people, gay people, physically and mentally afflicted and others that did not fit into Hitler's plan for that perfect Aryan race. Some of the accused are leaders in our world today. Even Kurt Waldheim, former United Nations Secretary General and Austrian head of state, has been accused. The Hitler men are touted as cruel, criminal and responsible for their acts even though they acted upon orders of the superiors. Some of these men, when found living in other countries under different names, have been executed for their war crimes. Yet, we find many people who follow orders of those perceived to be authorities even though there is absolutely no information to that effect. Do people want to be told what to do? Are people so afraid of unforseen consequences that they will succumb to any sort of pressure from anyone?

THE ROOM OF HORROR

A few years ago a department of a highly respected university in the United States did a study of reactions to authority. The results were astounding, sickening and frightening. That was my sentiment when I first read of these studies and it is my sentiment now. The only difference is that I now have several more years of experience in the world. One thing that this experience has done is to make me more aware that the results of these studies is exemplary of the United States social condition today. Below is a brief description of that study.
    A social scientist, wearing the kinds of clothing that you see worn by St. Elsewhere doctors and other television hospital people, opened an experimental office on the university campus. He recruited passersby to be a part of the experiment. The experiment was to observe the effects of electric current upon the human body. He showed the volunteers

a room where a person was hooked to an apparatus similar to an electric chair. The volunteers were told that they would be in another room where they would not be able to see the person receiving the electric current. However, a sound system between the two rooms would allow them to hear the movements and comments of the person receiving the electric current.

Once the volunteers saw the person being strapped into the electric chair type contraption, the door was closed. Then the volunteer was taken to the room where he would administer the electric shock. The volunteer was told that he should apply the current as the scientist instructed.

Have you heard of this experiment? If not, brace yourself because it happened in the United States with average working American citizens.

The scientist started the experiment by having the volunteer administer small amounts of electricity, maybe ten volts. Then the scientist told the volunteer to increase the amount gradually until the victim was receiving hundreds of volts. As the voltage increased the exclamations and grunts of the victim became more intense.

At points during the experiment a volunteer would ask if they should stop the experiment because the voltage was too high. Always the scientist insisted that the volunteer continue to apply the voltage. The scientist told the volunteer that he must complete the experiment.

As the gasps and groans of the victim became cries for help, the volunteer often hesitated, but at the nod of the scientist would apply more electricity. After several increases in the voltage the groaning victim would finally say, "I'm dying." After the victim, from all audio indications, had died, the experiment continued. The scientist had the volunteer continue to apply more and more

electricity, even thousands of volts.
    When the session was over the dead victim was fried as far as the volunteer knew. The volunteer had accepted the authority of a person he had never seen. He didn't really know the name of the scientist. Yet, at the scientist's command he seemed to be willing to kill someone.

Your first response is probably typical of those who have heard of the tragic example. "The volunteers that continued with higher and higher voltage were certainly in the minority." No, they were in the vast majority! In fact, only a few refused to follow the orders of the scientist and demanded to know the condition of the victim.
    The results of this experiment were so alarming that the experimenters tried to rationalize that the only reason so many of the volunteers would succumb to such unestablished authority was that it was on the university campus. They reasoned that volunteers willingly participated because they knew the university would not allow murderous experiments.
    The experiment was taken off campus and set up in a shabby storefront. New volunteers were recruited as they moved along the streets.

Brace yourself!

The results were the same. For no apparent reason volunteers accepted 'authority.' Hopefully this is not typical of our society, but it certainly has BEASTly implications. We can hope that as with the white glove fad, the authority studies are not an indication of the actions of the majority of the people. We hope that it is more difficult to draw most people into any fad, especially one as atrocious as the electric shock experiment.
        We can hope!
        Certainly, to have the total world

Sign Of The Times 61

population participate in anything, good or bad, it is necessary to reach or contact them. In years past that was impossible. There must be a very subtle and sophisticated mechanism for reaching the world population to implement the MARK.

Today, there is that subtle, sophisticated mechanism!

CHAPTER 5
FAULTLESS CONTROL

Over the years many systems were developed which appeared to be the mechanism needed by the Antichrist. Each of these systems had faults. However, with each new system, some faults of the older systems disappeared. Finally, today, there is a system which has eliminated almost all of the faults.

Early in the evolution of that system was the bar room re-entry ink stamp. It was a simple, but effective system in the control of unpaid customers. A person was at the door. If he did not have the stamp, he was not allowed to enter until he paid. Then he was given the stamp. He then could exit and re-enter by showing the stamp to the doorkeeper. This type system was necessary for clubs, bars, and other establishments requiring cover or door charges. The charges are usually for an entire evening or day. However, a customer or client may not desire to stay the entire period of time. To allow exit and re-entry it is necessary to identify the paid customers.

These re-entry ink stamps have been used for years, but there are many problems. The ink stamps smear and are unsightly for evening wear. But, more important for the use in the reign of the Antichrist is the fact that they are not permanent. They fade or wash away.

Some problems inherent with the re-entry ink stamp were overcome in the sixties and seventies with black lights and black light ink. A small black light was kept by the entrance of the bar or club. When paying, a person has his hand stamped with an ink barely visible to the naked eye. However, in the beam of a black light the ink stamp is very visible.

Although the black light ink seemed to be

the answer to the cosmetic problem of the re-entry ink stamp, there was still a problem. As with ink stamps, there was little stability. When washed the black light ink disappears.

In the late seventies another mechanism was developed. It seemed perfect for use during the Antichrist reign. This mechanism was an invisible tatoo substance. A tatoo with this substance can only be perceived or "seen" by laser rays. Even this mechanism has drawbacks which raise questions.

Probably the first question to come to mind is: "How will you read it?" Will you be put under a microscope or "laser" light every time you check out the groceries?

Almost! Certainly every time you check out the groceries you will pass by a laser reader. Your tatoo is readily perceived by laser rays.

Yes, laser rays are used for everything from destroying cancer cells to welding retinas in the eye. Now they are going to be used by the Beast.

The laser technology answers many of the questions about unsightly identifications stamped in the forehead. Yet, other questions remain. For example, how would you avoid the use of identical numbers?

Although it is a logical question, I believe that it is a moot issue. The years of pondering over a mechanism and a numbering system were all wasted. There will be only one number, the adoration code 666, as described in chapter 1. The individual will be identified through other high technology methods once the adoration number, 666, has been recognized. Those who feel a need to enhance God's meaning of 666 are reminded that when God speaks he means what he says. God does not need man to edit his inspired messages. The Creator of the human being built into man an identifying system while creating man years ago. Man is just beginning to use the identifying system that was there with Adam and Eve.

Faultless Control 65

FOOLPROOF IDENTIFICATION

The Adam and Eve identifying system is perfect for the Antichrist reign. It is foolproof. For man it is a new and intriguing technology, Biometrics. Biometrics includes the science of identifying unique physiological characteristics of a person. Once the characteristics are measured the person is forever identified.

Fingerprinting is a simple predecessor of Biometrics.

Two uniquely identifying features of Biometrics are:

    1) Retina (eye) blood vessel patterns,

    2) Special structures of sinews and blood vessels in the webbing between the fingers and between the toes.

These two identifiable patterns appear to be more unique to a person than fingerprints. In addition to being a better identifier, the process of identifying is not nearly so messy as fingerpinting. The readers look at the blood vessel patterns through the skin. Neither the skin nor the blood vessels are damaged regardless of the numbers of times they are read. In addition to the blood vessel reading technology there are other biometric developments that make the Beast identifying system more complete.

One interesting technology being developed is for identifying or verifying signatures. Throughout the years police and other investigative organizations have relied on handwriting experts. These handwriting experts are good, but, would you trust them to determine the fate of your own life? Recently a good example of handwriting used for identification was presented in worldwide news articles about dead Nazi war criminal, Mengele. In the articles there were two samples of handwriting purported to be of the same person. The two samples did not seem close to identical and yet were used to identify Mengele. Such comparisons will not be a problem with the new technology. It will be virtually foolproof.

The new technology incorporates two computing techniques, analog and digital. The two terms may be unfamiliar, but a general understanding is possible through fairly simple illustrations.

Analog computing is continuous. A simple example is a curve. If you draw a curve with a pencil there are no breaks from beginning to end:

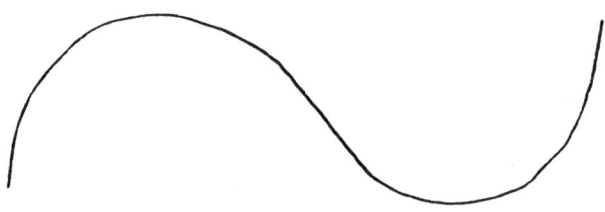

This kind of computing is used to measure things like pressure on a water line or the flow of the water or gas through pipes. The measurements are taken continuously and are often recorded on a chart or screen as a continuous line. You have probably seen other examples of this in medical equipment such as heart monitors or brain scans.

The word continuous describes analog and

Faultless Control 67

distinguishes analog from digital.
   Digital computing is said to be discrete. Discrete simply means that it is not continuous. Using the same curve for illustration we have:

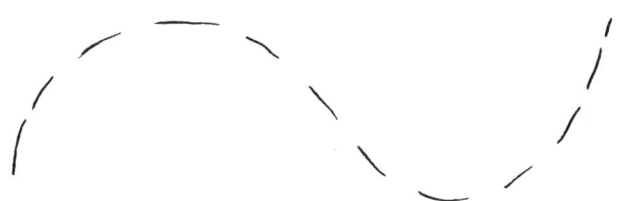

If you have not dealt with such thoughts before, you may jump to the conclusion that analog computing is the answer. However, due to the way the human eye works and due to the physical nature of matter (things), digital computing can easily appear to be analog. This principle of digital appearing to be analog is easily illustrated by photography and high level mathematics.
   A whole discipline of mathematics called Calculus is built upon this idea. If a curve is broken into enough smaller and smaller discrete parts it will approximate or appear like the true continuous form of the solid line drawn above. This concept is very important, but, before we get bogged down into a discussion of Calculus, let's do a simple experiment to see the effects of the digital curve appearing to be like the analog curve.
   Hold this book open so that you can see the discrete curve shown above. Next place the book on a table or chair so that you can see the curve. Now step back farther and farther away from the book. As you move you notice

that the curve begins to look more and more like a solid line.

Another good example of the discrete particles appearing to be contiuous is the movie film of a person walking. If you hold the film before a strong light you are able to see a discrete (one) photograph or frame. Each frame on the film strip is a separate photograph of the person in a slightly different position. As we move the film through a projector the person seems to be in continuous motion. The more frames that you have and the faster the film passes through the projector the smoother the motion of the person. It certainly is a change from the old movies of the nineteen thirties. The characters moved, but seemed to be always jerking from one position to the next. The speed of the discrete frames or pictures makes them apppear as a continuous or analog type movement.

These same principles are used in the modern technology for identifying signatures. The pressure sensitive detectors in an analog fashion follow the continuous lines of the signature. Not only are the lines followed, but the angle of the pen or pencil, the pressure applied, and the speed of the movement are measured and recorded. This continuous information is approximated into millions of bits of information and saved in the computer's bank of storage equipment.

With the analog-digital concepts and the two phases of biometrics discussed, we have covered practically all conceivable needs of personal identity. Yet, biometrics and other high technology areas are moving into other means of identifying, including voice.

When thinking in terms of rights of privacy of information, these biometric and laser technologies are frightening and disgusting. From a scientific standpoint they are intriguing. From the standpoint of the philosophy of the Antichrist these technologies are BEASTLY.

QUALITY OF LIFE

BEASTLY! Yes these technologies do seem Beastly, but they will not be brought into our lives by force. The people will beg for these dramatic improvements to the quality of life. These life qualities fall into general categories such as:

> SECURITY,
> COMFORT,
> EFFICIENCY,
> AND MENTAL STABILITY/PEACE OF MIND.

With the increased qualities of life we also see some negative points.

One disturbing point is the fact that new born babies will be subjected to the Mark with no choice in the matter. However, the same situation exists today. Children are born to child molesters, child abusers, financially inept parents, other children and many other atrocious situations over which they have no control.

Many 'right wing' Christians in the fight against abortion seem to forget that these situations exist. A pun I heard during the 1984 presidential elections somehow says it quite well, "According to one group, life begins with conception and ends with birth." It would be very interesting to see the results of a poll of those who, based on religious beliefs, support capital punishment and fight legal abortion.

Some things appear quite amusing if you can forget the atrocity to human beings. Consider a man like Senator Jesse Helms of North Carolina while fighting legislation for legal abortions, in 1985, relinquished the chairmanship of the prestigious Senate Foreign Relations Committee to remain over the Agriculture Committee.

Why did he refuse the more prominent committee leadership?

According to some, it was to be able to wield more influence in favor of R.J. Reynolds, Inc., a company known for its tobacco. If this is true, it certainly raises a good question, "Which kills more human beings - tobacco or abortions?"

For an answer to this question, ask the American Cancer Society, The American Diabetes Association, and The Heart and Lung Association. According to an article in the Charlotte Observer, October 29, 1986 30% of all cancer deaths are linked to smoking! In 1987 a 26 year study in Framingham, Massachusetts indicates that stroke is caused by smoking.

Because of these atrocious situations currently existing and the method of society to treat the symptoms rather than the causes, most of today's society will see the effects of these new technologies only as improvements to the quality of our lives.

THREE OR FOUR DIGITS

Consider the benefits of analog computing for signature identification. Today a person can forge a signature which eludes human scrutiny. But, using the analog/digital methods described on pages 66, 67 and 68, forgery will be eliminated. The signature data will be digitized and stored in the personal records just as the Social Security number would be stored. And, no one will be able to forge or duplicate the type information gathered.

To add to the affinity or usefulness of biometrics and computer technology in fulfilling the needs of the Mark, entrepreneurs are suggesting that the biometrics identification be further encoded with a prefix of three or four digits. This further encoding may indeed establish an identification code for which no one could decipher. Only the BEAST could decipher the code. The code would be of the greatest

Faultless Control 71

encryption. The three or four digits of a digital nature and the biometric portion of the analog nature. Specifically:

> 666 + your Biometric identifier – finger webbing, retinas, etc.

Immediately you begin thinking of ways to outsmart the BEAST. You think, "I'll just move to another state and escape him."

NO YOU WON'T!

Even if you move to Sri Lanka you will not escape him.
Escaping the Beast will not be easy!
In his book, 1984, George Orwell portrayed a Big Brother who would know your whereabouts at all times. You could do only what Big Brother wanted you to do. Well, the BEAST encompasses all that Orwell predicted and more. As my Father had no way of knowing how the BEAST would accomplish all that was prophesied, Orwell did not have the advantage of seeing the physical mechanism of Big Brother.
Today we have that opportunity. We see this fascinating technology every day. There is no need for the BEAST to wait any longer. The technology is here, the people have the expertise to put it into effect and --- Society Is Ready To Accept It!
On page 69, I said that improvements in the quality of life will cause people to readily accept the Mark. Those qualities were security, comfort, efficiency and mental stability. Those qualities of life are becoming more available with each new development in technology.
As stated before, one of the most important areas of development in technology over the past few years is the computer. One computer, the abacus, has been around for hundreds of years, but only in the last few decades has the computer technology advanced

at phenomenal rates. The progress chart below will give you an indication of the rapid developments.

TIME LINE

- 1823 Charles Babbage – begins work on machine to solve general algebra problems.

- 1835 Samuel Morse – telegraph message transmitted.

- 1876 Alexander Graham Bell – the telephone.

- 1885 Allan Marquand – designed his electric logic machine.

- 1890 Herman Hollerith – designs a tabulating machine. This is the same man for whom the Hollerith (IBM or computer) card was named. These cards have been used by everyone in such ways as: tax refund checks, social security checks and company checks.

- 1906 Radio Broadcast – the first.

- 1924 IBM – The company, Computing-Tabulating-Recording becomes International Business Machines.

- 1930 Claude Shannon – PhD thesis explains electrical switching circuits modelling Boolean logic, the basic idea of computer circuits.

- 1930 Kurt Godel – predicate calculus deduction systems.

1936	Benjamin Burack – builds electric logic machine.	
1936	Television first broadcast.	
1940	John Atanasoft and Clifford Berry design a vacuum tube computer.	
1941	Pearl Harbor attacked.	
1943-1946	John Mauchley, John Von Neumann and J. Presper Ekert build ENIAC, the first all electronic computer.	
1947	The transistor is perfected.	
1962	Stephen Wozniak – won local science fair prize for his addition/subtraction machine.	
1964	John Kemeny and Thomas Kurtz, Dartmouth College develop the first BASIC (Beginner's All-Purpose Symbolic Instruction Code), a programming language.	
1968	INTEL – integrated circuit manufacturer was founded.	
1971	Intel develops the 8008 chip.	
1971	Stephen Wozniak and Bill Fernandez build the Cream Soda Computer.	
1972	Gary Kildall writes PL/I, the first programming language for the 4004.	
1972	Stephen Jobs and Stephen Wozniak sell blue boxes, forerunner of Apple Computers.	

1973	Gary Kildall, Ben Cooper build their astrology forecasting machine.	
1974	Intel develops the 8080.	
1974	John Torode and Jerry Kildall begin selling a microcomputer and a disk operating system.	
1976	Michael Shrayer creates Electric Pencil.	
1977	Apple Computer opens its offices in Cupertino, CA.	
1977	Tandy/Radio Shack announces first TRS-80 microcomputer.	
1981	Osborne Computer Corporation incorporates and introduces the first portable computer.	
1981	IBM announces its personal computer.	
1982	Digital Equipment Corp. (DEC) announces personal computers.	
1982	Apple Computer announces new small computer, LISA.	
1982-1986	Human brain grafts research.	
1983	IBM announces a smaller personal computer, the PCjr.	
1984	Apple Computer announces a smaller, more versatile computer, the Macintosh.	
1984	Biochip research.	
1985	Chip credit card.	

Faultless Control 75

1986   Optical computing research.

1986   "Star Wars" technology debated.

1987   Texas Instruments introduces world's first artificial intelligence microprocessor on a single chip.

1987   Ape-Human. Embryo created. Human father. Ape mother.

From this list we see how rapidly the computer industry has developed in the last few years. It is interesting to note the entry of IBM into the little computer industry. For the first few years it was little, unknown companies that built small computers, but it soon became apparent to everyone, including IBM, that little computers are becoming an integral part of our lives. They are also the heart of any Antichrist or Big Brother type mechanism.

Computers, biometrics and other technologies combined will create a Beast which is beautiful, kind, and charitable. Attitudes will change as more is learned about his hypocrisy and shrewdness.

YOU SUPPORT THE BEAST!

When you finish this book, you will either shudder at the mere mention of the words high technology and computer or you will have developed such a keen interest that you will run out and buy your own personal computer and become an innocent supporter of the BEAST.

An innocent supporter of the Beast?

How can you innocently support the Beast?

Should we not avoid using the mechanism of the Beast?

These are questions that only you can answer! However, if the Supreme Being sent the message of the Beast and the Antichrist through chosen, obedient prophets, it seems that it is not our role to stop it. Our role is to continue working, and preparing for the return of the Messiah.
 Then why be afraid to follow the intrigue of modern technology? Human nature is to follow the intrigue.
 An admonition for the fundamental believer may take away the worry of using the Beast mechanism. "Live right! Be prepared when Christ returns to take his saints away!"
 According to Matthew 24:22 and Revelation 3:10, if you follow the admonition and if the fundamental teachings are correct you will be taken off the surface of the Earth prior to the reign of the damnable Antichrist. Thus, "living right" is the only way to avoid the horrors of the Antichrist. Using his equipment will not doom your soul. All of human progress may be considered preparation for the Antichrist. Without electricity the computer would be of no use to the Beast. Radio and telephones are steps toward the Beastly communications. All progress unwittingly aids the Beast.
 In 1823 when Charles Babbage began to build the "world's first computer," neither he nor anyone else assumed that it would be the machine of the Antichrist. Did Augusta Ada Byron, the first computer programmer, have any idea that her programs were aiding the Antichrist? Probably not! In fact many probably saw the computer as the toy of a rich kid. After all was Babbage not a rich kid, the son of the Lady courting Lord Byron, the father of Augusta Ada Byron?
 Although Lord Byron may have been the father of Augusta Ada and the sugar daddy of Babbage, he was not the sugar daddy of U.S.

Faultless Control 77

corporations. However, as late as the fifties, computers were still considered the toys of big corporations. This was true mainly because it took so much effort and money to maintain a computer. None other than wealthy kids and large corporations could afford one.

It took such massive amounts of space to house computer systems. The Central Processing Unit (CPU) and it's peripheral equipment needed to be housed in a well air conditioned room. There were many vacuum tubes which needed cooling. Even then the efficiency was not acceptable. Fortunately some people kept playing! With the advent of the transistor the efficiency was drastically improved.

Now a small desk top personal computer can do the same amount of work that the largest computers could do in the sixties.

THE SILICON WALL

In the seventies chip technology became the buzz words.

Chips were usually distinguished by numbers like 8080, 8008, 4004 or 80286 (see INTEL on TIME LINE on page 72). The chip is basically a tiny piece of silicon onto which circuits are printed or etched. These chips are used for software (programs) or storage (information). A piece of silicon the size of your thumbnail may contain volumes of data or programs.

Now, in the latter half of the eighties silicon chips may be nearing obsolescence.

Technology and silicon seemed to have reached a maximum, sometimes referred to as the Silicon Wall. Some of the circuits on these chips are smaller than bacteria. We have done about as much with the silicon chip as is possible.

However, at least one company is shooting for the Biochip. The biochip will have one billion (1,000,000,000) times the capacity of the silicon chip. This means that several complete libraries can be stored (Washington Post Magazine 1/6/1984) on one chip the size of your thumbnail!

Another technology being researched at AT&T Bell Laboratories is optical computing. Instead of using electrical systems, this technology will use light signals. This technology is predicted to make almost a trillion actions per second. We could call these teracts per second. Some of these little chips may contain more than 10,000 tiny mirrors. Some predictions are that all five billion people on earth could be talking on the phone at once using one of these tiny chips (Business Week 7/28/86 p 48 "Pushing Computers closer to the Ultimate Speed Limit").

Although the earlier computers were large and cumbersome, they were a start. They gave my father the facts needed to understand that we were approaching a machine that could maintain all of the information the BEAST would need.

Today it is obvious. The technology for the Beast is here. Yet, one persistent question remains. How will the BEAST dupe the people into accepting such an ominous bill of fare?

Faultless Control

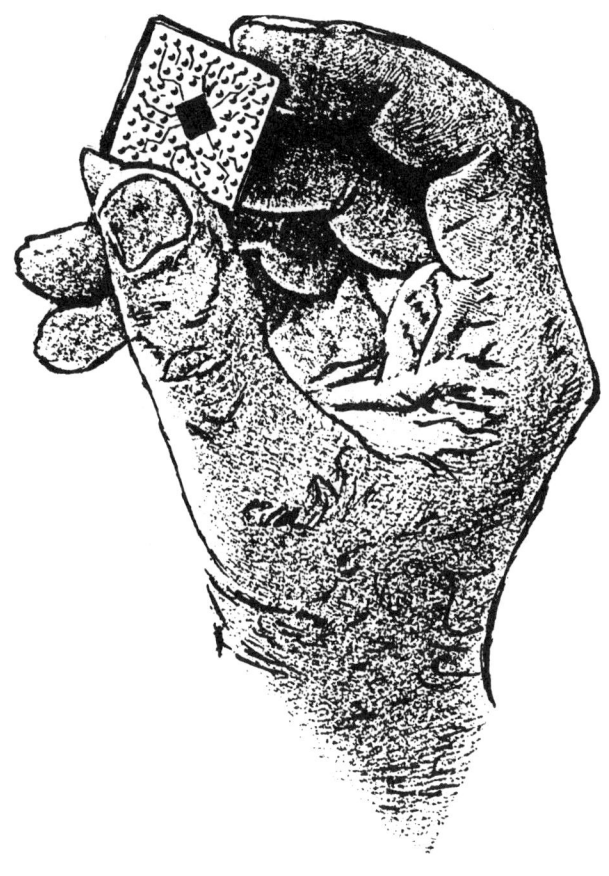

CHAPTER 6
HANDSOME ANTICHRIST GIVES THE GOOD LIFE

The Antichrist bill of fare becomes less ominous or frightening if presented properly. The Antichrist will present it as pretty and helpful. It will be presented as an aid to the quality of our lives, not government controls and restrictions. When presented in this manner there is no longer fear. Security and convenience are two of the life qualities which the Beast will use as selling features. The foundation for this approach has already been laid.

Years ago a mechanism of controls and record keeping began. It began to improve our quality of life. It was presented as a way to provide every American with a retirement income. It was never presented as information gathering and control. Yet, the Social Security mechanism has become that today. At sometime in the future it may be only a record keeping agency. Today there is certainly a question as to whether the Social Security system will eventually discontinue the retirement income for the American citizen. Whether or not the retirement income for the elderly continues, there are few among us who fear Social Security enough to run from the postman when the Social Security check arrives!

However, since the mid-nineteen thirties it has caused some concern for some fundamental believers. Early in its existence the Social Security number certainly appeared to be a good predecessor of the Mark. In any event it started a trend which is still with us.

From the thirties to the eighties the FDR salvation number, the Social Security Number, has evolved from a retirement benefit to an all encompassing charity. It has grown very cumbersome and the actual benefits seem to be dying because it has become such a costly program for the government and allegedly the funds are used for other things. It has also become a major political hot potato.

In the 1976 presidential forums and debates, candidates were asked if millionaires should receive Social Security checks. Two elections later, in 1984, the issue was not so much, "Who should receive the checks," but, "Will there be any checks? Should the program continue?" After the 1984 election was over major cuts in the Social Security payments to the retired was the hot issue.

The hot political issue of "WHO GETS PAID" has taken the attention away from the issue of most importance to the Antichrist. That issue is record keeping. Attention or no attention, record keeping is a big part of the Social Security program. Almost every citizen is innocently involved in this record keeping. However, as late as the mid-sixties there were some who managed without a Social Security number.

One college friend of mine did not have a Social Security number. He said that no one in his family had a number. He said that as a Roman Catholic he should not take a Social Security number. He believed it was evil. I was surprised that a Roman Catholic would believe this way. However, it would have been no surprise at all if the same statement had come from someone in the Assemblies of God or the Pentacostal Holiness Church. No Roman Catholic that I have known since that time has explained the basis of the beliefs of my college friend.

Today, it would be hard to survive without

the number. Even the university which we were attending at the time now uses the Social Security number as the student identification number. Further acceptance of the number as a universal identification was recognized when the armed services began to use it as the identification number for servicemen.

Was this another flag raised for the fundamental believer? Is it more obvious that it is part of the mechanism of the Antichrist? Now many corporations and other organizations use the Social Security number as the employee or member identification number. As it is gradually being accepted, there is little thought of it being ominous. The Beast does not need to bully anyone into accepting his useful mechanisms. They will improve our quality of life!

It makes good sense to use the Social Security number for everything. From a business viewpoint it's a time and manpower saver. It's easier for the individual to remember only one number. For the organizations it means less computer space.

The Social Security number is here to stay for a while. It may even change names but its progeny or offsprings are here forever. The Social Security number is in the United States, but other countries use similar systems.

It would be convenient and practical for all countries to combine their systems for a worldwide number. Currently, there is some effort to make that a reality. In a few years we will likely intersperse our numbers with numbers of other countries into a common system. As a world identification system develops, Biometric technology will probably be used.

There already exists in the minds of many a need for this kind of international system. Look at the travelers! They need an international identification system. Such a system would benefit travel agents, airlines, restaurants, banks and many other businesses, but most of all it would benefit the individual traveler, you and me!

The stage has already been set for that too:

"Plastic Money" or credit cards!

Once upon a time most of the people around you had cash? But, today credit cards are almost as common. In fact, cash is not accepted in some business transactions such as renting a car from one of the major car rental companies!
　　Just project into the future. Would it not be easier if each person had just one number for all charge cards whether the cards are from Chase Manhattan Bank, Sears, American Express, First Interstate Bank, or Eastern Airlines.

In 1985 a company announced that it was going to make a chip card. That card would have a storage capacity of 8,000 bytes. A byte is usually one character, number or letter. The card would look much like the ones carried by millions of Americans today. The chip card would not have the magnetic strip across the back. With the chip card you would probably have only one card. That one card would have all of your card numbers on it. Eventually this type card could contain all of your vital information, from driver's license number to credit card number to blood type.
　　Although the Social Security number, used universally, makes sense from an efficiency standpoint, it may seem outdated once Biometric identification is perfected.

Biometric identification is certainly a more sophisticated approach. However, for that sophistication there is a great price. That price takes many forms. The first form is that of privacy for individuals. Another form is sheer cost. The cost for the computer storage of twenty digit numbers is much less than the cost to store blood vessel patterns. However, this cost factor would be practically eliminated if Biochip technology is perfected. If you recall from earlier mention of the Biochip, the capacity for storing information will be one billion times that of a silicon chip. Some predictions are that this technology will be available by the year 2,000.

Perfect timing for the prophecies!
    Another reason that the Social Security number may remain for years to come is transition time. This is the time it takes to change from one system to another. Often the transition time is not dependent upon the physical change, but upon the time it takes to change the mental attitudes of the people to be affected by the change. In this case it may be the time for a generation to come of age. This would be about twenty years and would put us near the beginning of the twenty first century. This is very significant timing for the Antichrist. Many people speculate that this is the approximate time for the next major event of Bible history and prophesy.

Approximate Year	Event
0000 or 4000 B.C.	Earth's creation
2000 or 2000 B.C.	Abraham
4000 or 0000	Christ
6000 or 2000 A.D.	Antichrist

It is a perfect calendar event for the Antichrist and it allows time for that mental conditioning or transition time.
    The transition time from no system to the Biometric system would have seemed forever. It may have appeared more ominous. However,

everyone has grown accustomed to being referred to by number. If you are over forty, you remember when you had a different number for each function of your life. If you're under twenty five, you may have trouble understanding why all of the numbers were ever necessary. The likelihood is that you have had only one number - your Social Security number.

But, remembering only 666 is even better!

With the 666 adoration code and your Biometric patterns, eventually you will not need to take up brain space remembering personal identification numbers.

For Biometric identification to be effective worldwide, three things must happen:

- --The social climate must be right for acceptance of the system of identification by the people.

- --The massive task of collecting all of this data for every individual on the face of the earth must be completed.

- --A method must be developed to create, maintain, and provide access to the data - a massive Information Center.

These three conditions are very near!

Consider the last one. Almost every major corporation has an Information Center. The TRW Credit Data System and the Social Security System collection of names and numbers are examples. In fact each person reading this is very familiar with an information center without being aware of it.

The term Information Center probably conjures up thoughts of a computer to many persons. In fact, that is what many conceive a computer to be. In many ways that is the right concept. Recently, however, the term

Information Center has been expanded to a meaning of greater breadth. It now refers to a body of computerized information that may be used by several persons. The data is stored somewhere in a massive computer system. All persons who need that data may retrieve (get) it from that central location and massage (arrange) it to meet the individual need.

An example of this is accounting information. A large corporation with many subsidiary companies spread across the country may have each store processing a credit application for you. If the corporation has only one large database combining customer information for all stores, the need for more than one credit check would be eliminated. Once all the data was stored and each subsidiary store had access (via a small terminal) to that data base, the store would need only to type in the customer name and Social Security number and the response, good or bad, would be reflected on the screen or the typewriter like printer. This can be done in seconds regardless of the location of the central computer. The computers talk to each other and they talk fast.

A good example of these fast talking computers is found in the airline industry.

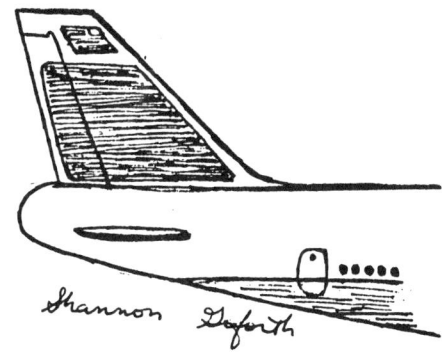

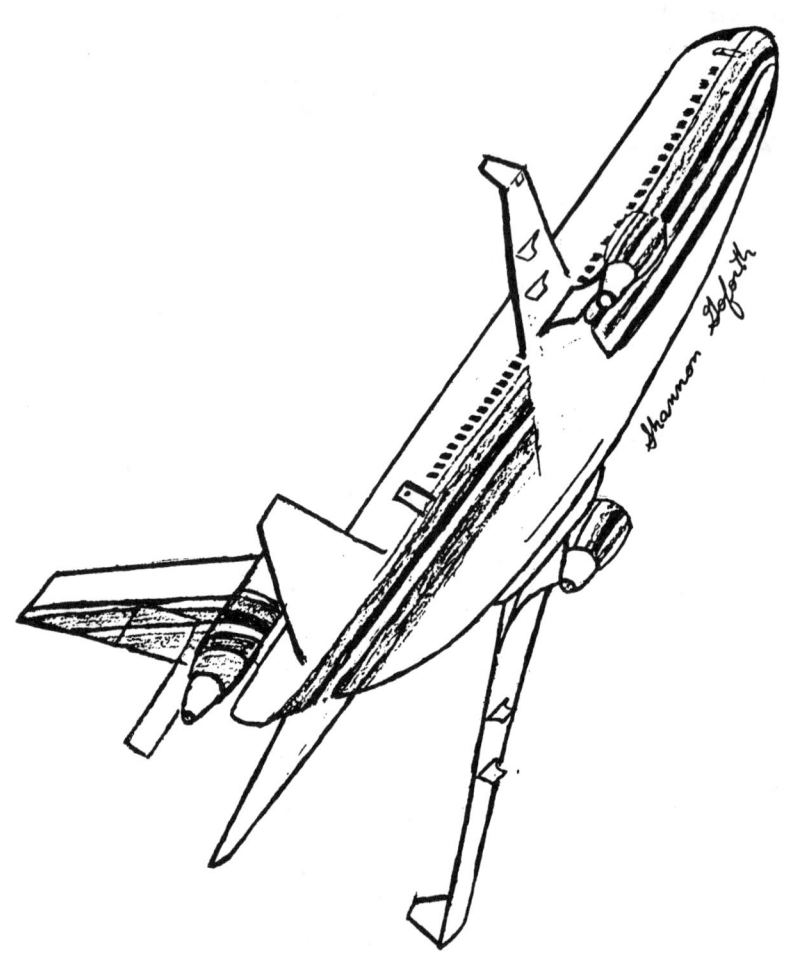

CHAPTER 7
ANTICHRIST AND THE AIRLINES

The airline industry was one of the first groups of business organizations to make extensive use of high powered computers and sophisticated communications equipment. The first major use was for seat inventory. Each airplane had a certain number of seats for sale between cities on the flight schedule. All seats for every flight of the airline were counted and marked off as reservations were made. The seat for a particular flight was 'sold.'

To control inventory the airline needed its ticket and reservations locations connected to the computer by telephone or other communications lines. This was a great endeavor, but a great enhancement for the airline operation.

Before computers, the reservations system was primitive. Records were kept on chalk boards. When an agent answered the telephone and a seat was sold, a mark was put beside the flight number on the chalk board. Stations or airports downline (i.e., the next airport city on the schedule) were called. This has changed dramatically.

Now an airline with its computer base in Miami, Florida (MIA) supplies information to its agents all over the country from that one location. Other airlines have their central computer systems in Atlanta (ATL), Los Angeles (LAX), Tulsa (TUL), Rockleigh, N.J. (RNJ), Denver (DEN), and Kansas City (MKC). Each of these computer systems performs basically the same function for the respective airline. And, the computers talk to each other.

The airline computer system is a mark of great human ingenuity and a lunge toward the Antichrist mechanism as the following paragraphs depict.

Suppose you want to take a trip from Richmond, Virginia(RIC) to Anchorage, Alaska(ANC). On the way you would like to stop in Los Angeles, California.

Find Airline A, an example airline, in your telephone directory and dial the number.

The agent who answers is probably in Charlotte, N.C. If not in Charlotte, the agent is in any one of seven other reservations offices across the United States. If you are thinking, "Only eight reservations offices for a major airline," you are right. There is one in Charlotte, Atlanta, Miami, Chicago, New York, Houston, Los Angeles, and Salt Lake City. If you call after midnight you are almost surely talking with an agent in Charlotte.

The telephone system automatically searches the network of reservations offices to determine the office with the least workload to handle your call. This telephone system is pretty fascinating, but compared to the total airline computer and communication system it is minor.

When the agent answers, you tell him where you want to go. Be sure to include the stopover in Los Angeles.

When the agent checks the itinerary you have given him, the following response may be typical:

> "Sorry to keep you waiting. I have your flights. You will fly on Airline A from Richmond to Atlanta where you will connect to Airline B to Los Angeles. After three days in Los Angeles you fly Airline C to Seattle and connect with Airline D to Anchorage."

You sit dumfounded. The agent made your reservations to Anchorage in only a few seconds. Not only has the agent confirmed a flight on Airline A, but also on Airlines B, C and D.

How did an agent in Charlotte do this?

The reservations computer for Airline A is in Miami.
 Yes, you called a number in Richmond and reached an agent in Charlotte who, sitting at a television like computer terminal, accessed information in a computer in Miami, Florida.
 Suppose, the week after you made your reservation from Richmond to Anchorage you wanted to ensure that you still have your reservation. You are in Washington, D.C. That's fine. Just give airline A a call in Washington. If you really want to be tacky, call another airline and let them find out for you. When the agent answers, you ask her to confirm your reservation. She will ask you for the flight number, the date of travel and your name. As you answer, she will type the flight number, the date of travel and your name in the following format:

 0000/01JAN-Lastname/I

Instantly your complete record is displayed on her terminal.
 You may think:
 "So the agent accessed the computer of Airline A in Miami; but, what about the other airlines? Does Airline A keep all of the information about the other airlines?"

That's a good question.

The answer is, "No!"
 Although some airlines maintain computer

systems for other airlines this is not the case between Airlines A,B,C, and D.

The computer center for Airline A is in Miami, for Airline B it is in Atlanta, and for Airline C it is in Los Angeles.

Assume that Airline D does not have its own computer and uses the computer of Airline E in Tulsa, Oklahoma. And that is not all!

Another part of the total system is ARINC, Air Radio Incorporated.

ARINC, in Maryland, acts as a traffic director or mail sorter for the airline computers. Its computer and communications system is subscribed to by most major airlines and many small airlines.

So we have:

>Airline A computer center- Miami,
>Airline B computer center-Atlanta,
>Airline C computer center-Los Angeles,
>Airline E computer center-Tulsa,
>Airline D headquartered in Alaska,
>ARINC in Maryland,
>The agent in Charlotte,
>And you, in Richmond, Virginia.

Mesmerized? Here is what happened:

You called Airline A in Charlotte.

You told the agent your desired trip plans. She immediately checked for an airplane seat for you. She did this by typing the information into a CRT, Cathode Ray Tube. The CRT looks very much like a television screen. It is used as a terminal to display information from the computer. The CRT has a typewriter like keyboard which allows the agent to communicate with the computer.

The CRT is sometimes called a VDT, Video Display Terminal. For these television type screens, VDT may be more correct as it covers

Antichrist And The Airlines

all types of technology. CRT refers to only one, but you are safe to use VDT or CRT.

The agent typed the legs of flight for your trip and requested the computer to confirm seats on the flights for you.

AIRLINE	FLIGHT LEG	FROM CITY	TO CITY
A	1	RIC	ATL
B	2	ATL	LAX
C	3	LAX	SEA
D	4	SEA	ANC

In an expected response time of 15 seconds or so, the computer came back with the results described above. The question is, "How did the computer in Miami get all the information?"

First the computer checked seat inventory of its own Airline A. Only one leg of the itinerary, Richmond to Atlanta, was available on Airline A.

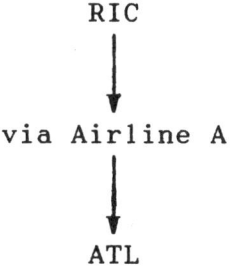

Now the real intrigue begins!

Learning that Airline A could supply only one leg of the itinerary, the computer checked with other airlines.

The Airline A computer searches its information to decide which airlines have the flights you requested to Alaska. The computer determines that Airline B has a flight from Atlanta to Los Angeles, Airline C has a flight from Los Angeles to Seattle, and Airline D has a flight from Seattle to Anchorage. But, to get seat inventory for Airline D the computer of Airline E must be contacted.

The Airline A computer does not always talk to the other airline computers directly. Sometimes it must talk through an intermediary like ARINC.

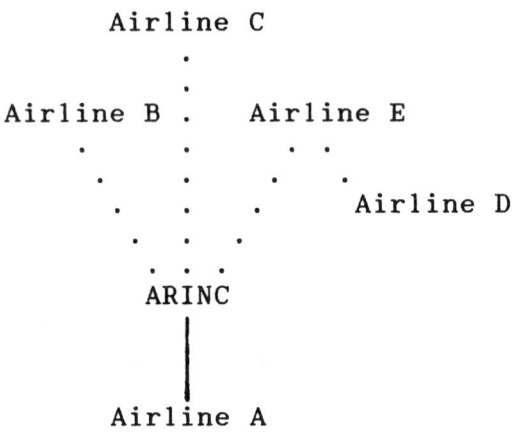

Airline A computer prepares messages or requests for the other airline computers.

Once the messages are prepared, Airline A computer puts addresses on these messages. Usually these addresses consist of three letters for a city code and two or four letters for the particular office code. Examples of city codes are:

    CLT for Charlotte, N.C.
    ATL for Atlanta, Ga.
    CHS for Charleston, S.C.
    CRW for Charleston, W. VA.

Once the messages are addressed, the computer

Antichrist And The Airlines

of Airline A sends the messages to ARINC. ARINC sorts the messages and sends them to the addressed airlines.

Almost immediately Airlines A,C, and E send answers to Airline A via ARINC.

Now the Airline A computer responds to the agent who asks you when you want to return from Anchorage to Charlotte.

You say, "January third."

The agent asks, "Do you want to stop at any city between Anchorage and Charlotte?"

You answer, "No."

The agent merely depresses one button and says, "Your reservations from Anchorage to Charlotte are confirmed." She then reads your whole itinerary. It takes longer to read the itinerary than it took to make the reservations.

The whole process took just a few minutes. An agent typically answers 120 calls in seven hours and twenty minutes (an eight hour day excluding break time).

It is incredible!

But, for the subject of the Antichrist, one thing makes this more incredible. All of this took place with NO HUMAN INTERVENTION except the agent typing your request into the computer.

Furthering these unbelieveable uses of the computer, let's continue until your reservation is completed and filed away.

The agent may ask when and where you plan to get your tickets. She may offer two ways of obtaining your tickets: tickets by mail and picking up tickets at a ticket office.

If you choose to get your tickets at a ticket office the agent will give you a time limit for getting the tickets. Usually the time limit is about half way between the date you made the reservations and the date you will be flying. When the time limit is set, the agent says, "Thank you for calling Airline A. Have a good flight."

As fast as this happened, something else will happen.

On the time limit date and time your reservations will be cancelled if you have not purchased the tickets. If they are cancelled, do not get upset with the agent. The agent did not cancel your reservations. The computer, again WITHOUT HUMAN INTERVENTION, checked the time limit, found that you had not purchased the tickets, and automatically cancelled your reservations.

Does this computer and communications system seem to be the baby Beast, an Antichrist perfect tool? Did it cancel the other airline flights too?

Yes! When Airline A computer checked and found that your ticketing time limit had expired, it cancelled your flight on Airline A and sent cancellation messages to Airline B, C and D at Airline E computer.

Again, all of this took place WITHOUT HUMAN INTERVENTION!

Beastly?

The Beastly part is the fact that the computer did it all. NO HUMAN INTERVENTION emphasizes the lack of the need for a harsh, demanding Beast. It just happened! The computer did it, not the Antichrist. As it develops more and more the Beast mechanism will be responsible for your being restricted in many areas. But, the many advantages will keep our minds off the negative aspects of the system. Another such advantage is the second way of obtaining your tickets. You simply ask the agent to send the tickets by U.S. mail.

All that you need to give the agent is your address and credit card number. The agent will thank you and in a few days you will receive your tickets.

Antichrist And The Airlines 97

Now your reservations are complete.

When you hang up the phone the thing that perplexes you the most is why the agent ever said, "Sorry to keep you waiting."

You were on the phone only two or three minutes. You sit amazed as you mentally traverse the path the message had to take.

We have not yet covered all of the details of the process. There are thousands of parts to the airline computer system and we will not attempt to cover them all. However, there are a few more parts that seem extremely important in the Antichrist mechanism.

First, if you had any thought of giving the agent a fake credit card number, forget it. Another task of the computer is to check for "black listed" credit card numbers. This protects the airlines from unscrupulous persons. Many of the airline companies have fraud prevention systems. ARINC also has loss prevention and fraud prevention systems which are subscribed to by airlines and other companies which accept credit cards as payment.

Another interesting factor in having tickets mailed is the ease of the process the second time you use the "tickets by mail" system. If you have used your credit card and had tickets mailed to you in the past, the agent may simply type in your telephone number. When the telephone number is entered, your address and your credit card number is displayed to the agent. The information which is displayed is appended to your reservations that were just created.

Is that not truly a convenience for you and the agent? It is also a money saver for the airline. Consider an airline which has a total of four thousand employees working in any one day. Assume that each agent saves only three minutes per day. That is a savings of twelve thousand minutes each day. That amounts to two hundred man hours or more than a month of one employee's time saved in one day.

The airline example illustrates the intricacies of the computer. It emphasizes the miraculous ability to instantaneously transfer information around the world.

Rapid information transfer was once restricted to one continent such as North America. This restriction was due to confinements of transmission lines. Those confinements have been removed by the use of cables in the oceans and with the advent of satellite transmission. Today you can easily have computers in London, England (LON) talking to computers in New York City, Charlotte, or Miami.

The technology is good for everyone and it is perfect for the Beast.

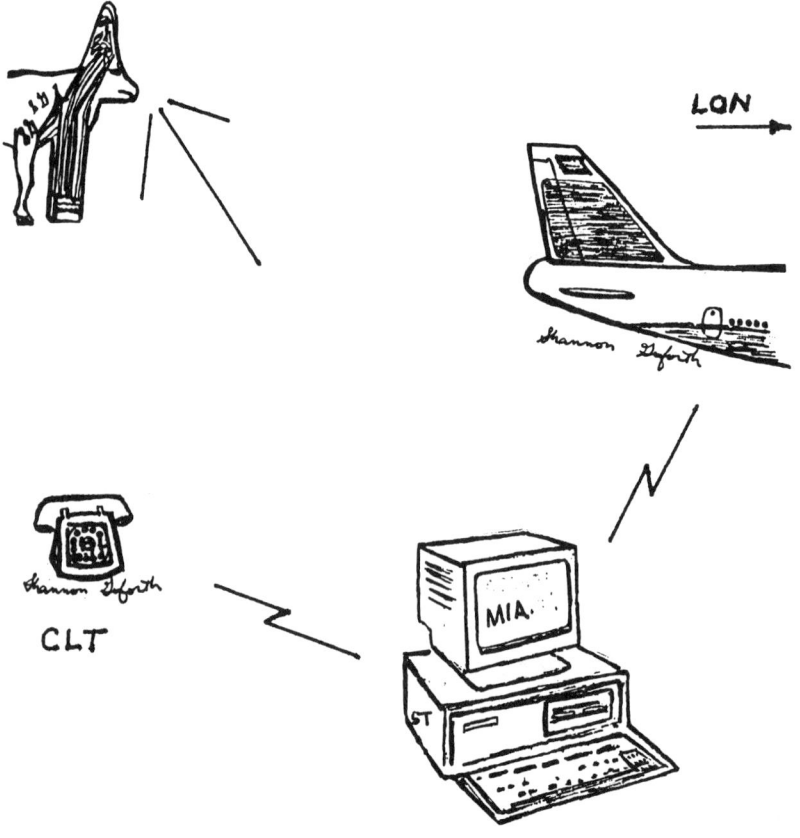

CHAPTER 8
COMPUTER TALK EASIER

In the late sixties the information systems were only available to large organizations like the airlines because they were so expensive. The computer alone could cost millions, and that was only the beginning. There were special building considerations, and specialized personnel. Extremely powerful air conditioning systems were required to maintain necessary temperatures. To ensure good circulation and ease of laying the electrical cables, computer rooms were designed with raised flooring. Clean room environments free of dust and smoke were mandatory. Computer rooms were probably more dirt free than many hospitals.

The employees were extremely technical. A computer is an intricate and valuable resource, but worthless if there is no one to program it. A program is a set of instructions, words or symbols, which tell the computer what to do. The person who writes programs is called a programmer.

In addition to the cost of the clean room, the computer, and the programmer, there is the cost of storage space. The computer storage space such as magnetic disk drives and magnetic tape drives was very costly. This required that the programming be done in a machine language which the computer understands. Later, easier to use languages were created. These languages were assembly languages. These languages are easily converted into machine language, but are still far removed from the normal English language.

An example of an instruction from one of the assembly languages is:

    B   TAG

B, the first letter of the word BRANCH tells the computer to go to another part of the program,TAG. Another example of the same instruction from another language is:

    JP TAG

The JP is from the word JUMP.
    These languages take advantage of the scarce storage space. JP is one half the letters in the word JUMP and the letter B is certainly a space saver over the word BRANCH.
    For years space saving was the greatest concern of the management of computer centers. Later the question of cost to train in the assembly languages versus the cost of the scarce space led to the use of higher level languages such as COBOL (COmmon Business Oriented Language). These languages are sometimes called third generation level languages because they are one generation later and one step closer to the English language than the assembly languages. The instruction to do the same thing as in the B or JP instruction shown above would be:

    GOTO   TAG

The GOTO is simply the statement Go To. In plain English you are telling the computer to go to the location TAG to continue working.
    Even COBOL, usually pronounced CO-BALL, requires extensive training and expertise to produce efficient software (programs). But, as the prices of computer systems decrease and the power of the computer increases, languages are being developed more and more like human languages. Naturally they are being called fourth generation languages.

Computer Talk Easier

Are you now beginning to see the Beast raising his head?

The use of computers is becoming easier and more widespread. The airline computer illustrates the efficacies of computers in our lives. But, this use of the computer only affects those who fly. Many people have never flown.

Many other industries have joined the computer revolution. Each new entrant brings the computer closer to us. The Beast creeps closer and closer in the guise of new innovations and improvements to our lives. The truth is that we all can see benefits of these electronic helpers regardless of our religious beliefs. Sometimes it seems that people readily see the advances in technology as BEASTly, but are unable to see the degeneration of social conditions as creating a perfect stage for the Beast.

DEGENERATION OF SOCIAL CONDITIONS

The degeneration of social conditions in our society makes use of the latest technology mandatory for security systems. These systems must provide security against acts of theft as well as security against slothfulness and ignorance on the part of the employees. One system that is very poignant in the discussion of information and surveillance computer systems is used by many grocery store chains today. The system has been around for many years, but only recently has it begun to be used extensively with optical character recognition (OCR) devices.

For an example of this, look on the next can of Coke you purchase. Near the bottom of the can you will find a rectangular area filled with lines and dashes. That is an encoded area so that the pricing and other types of information about packaging can be used. Almost all food packaging companies use similar forms of encoding. The strange little marks and the strange looking numbers on your checks can be read by a device called an optical scanner. This is the process used to tally your grocery bill when the cashier drags an item across a glass plate in the counter. The markings form a part of an identifying system using codes such as the Universal Product Codes. One day it is likely that everything you purchase will bear one of these codes.

Do you see the Beast?

Would you see the Beast if you had grown up in a church environment that taught these things about the Mark of the Beast, the Antichrist, and the final judgement?
   It certainly lends itself to the concept of the Beast!

ORGANIZATIONS, GOVERNMENTS, AND THE BEAST?

It sometimes appears that large organizations and governments are trying to help the Beast stand up?
   However, it is not an effort to help the Beast. Computers are labor savers and productivity enhancers. Computers do not make human errors. Computers can calculate your grocery bill much faster than a person with an adding machine; and, computers can figure taxes while you sleep.
   From the Beast standpoint the computer seems a bad omen, but from a business standpoint it is a godsend.
   Today the cost problem is diminishing rapidly. Think of the prices of watches and calculators at the nearby discount house. Today you pay three or four dollars for what you paid hundreds just ten years ago. Everyday we see indications that computer prices may continue to change. It is evidenced by computer company price wars such as Tandon's 1986 announcement of a new small computer

cheaper than the IBM PC and probably just as powerful.

Drastic price reductions began in the seventies. The changes became possible when tiny silicon chips and microprocessors were developed. As everything else skyrocketed through the double digit inflation years, computer power and storage became cheaper. These lower prices promoted the spread of computers. Home computers flourished for business purposes and for recreation.

The next time you have the opportunity to play Pac-Man, watch the little beast running over the screen and think back to the days when baseball was the greatest thrill of children. Think of all the time and effort it took someone to program the system to get these little characters to run around the screen gobbling up each other. Think how this one thing has changed the recreation in the United States in the past few months. Think of the money spent on this type of recreation. Ponder the psychological effects of this type of recreation on the development of our children. Is it good? Many argue, "Yes." Others argue, "No!"

In his book, THE DUNGEON MASTERS, William Dear, paints the main character as a gifted sixteen year old who is somewhat withdrawn from his college mates. Yes, the main character was in college at sixteen. The favorite game of this sixteen year old is, Dungeons and Dragons. The book is based on the life and suicide of sixteen year old, James Dallas Egbert III. The author, reportedly a flamboyant Texas detective, was hired by the boy's parents to find the boy after his disappearance in 1979. Detective Dear found him. One year later the boy committed suicide. In the book Dear portrays the boy as being obsessed with the computer game to the point that he actually tried to live out the game.

Such a story certainly makes parents think. Yet, are we to blame the computer for a disturbed child? Are we to blame his parents:

A domineering mother and a very shy, busy optometrist father? Or, is this just a "sign of the times?"

Psychologists and sociologists will probably argue these points for years. Their conclusions will probably be as sound as conclusions of other experts! According to Christopher Cerf and Victor Navasky in their book, THE EXPERTS SPEAK, Dr. Albert Einstein said in the early thirties that there was not the slightest indication that nuclear energy would ever be obtainable because it would mean that the atom would have to be shattered at will. We all know that almost a decade later Dr. Einstein with others had created the energy he said could not be obtained.

As we wait for the experts to make their profound conclusions we will probably come to our own. Whatever they are, pertaining to the effects of computers upon the young, we will all probably admit that they will make our lives better. Although the computers may be a signal that the Beast is nearby, we will use them more and more each day.

This book was written and edited on a personal computer.

Why?

I type very poorly. My small computer with word processing software has saved me hundreds of typing and editing hours.

CHAPTER 9
THE BEAUTIFUL UBIQUITOUS BEAST

HE'S EVERYWHERE, HE'S EVERYWHERE!

Whether typing or making airline reservations, the Beast mechanism will improve the quality of your life. The Antichrist will be welcomed by the masses. The masses will not see the Beast growling as he subtly raises his beautiful head. He will arrive as people look on, but do not recognize his true character. People will recognize him as nothing more than a benefit to our quality of life.

You may ask, "How will the Beast get us to accept the damnable Mark?"

It seems that it would be very difficult; however, it may be very simple. How many kids would readily accept a stamp of 666 in their foreheads for an indefinite number of free games of Pac Man, free tickets to a Michael Jackson concert, or free admission to a major league ball game. As witnessed in the "One Hand White Glove" fad there are many who would readily and willingly accept these tradeoffs, even though the tradeoffs will probably not be so blatantly and openly offered. They are probably going to be offered in a much more subtle manner than we would first imagine.

The ominous nature is gradually fading. Many people think the idea of the Antichrist is very silly. It is for the ignorant. It is often used in jokes. It is used in many non religious writings such as the book, <u>666</u>, by Jay Anson. It is often used in a very light and almost non suggestive way, but with a very innocent impact. On February 10, 1986 the television program, "Scarecrow and Mrs. King," was very interesting from the standpoint of the Mark of the Beast and the Antichrist theory. Usually this program is my light entertainment. For me it is a "who done it" which takes very little intelligence to solve. It, like most television programming, appeals to me only when I am extremely tired and do not really want to think. Well, February 10, 1986 was one of those evenings. I watched "Scarecrow and Mrs. King."

In the search for the bad guys Scarecrow and Mrs. King involved a friend who was a computer whiz. After searching many accounts they finally found the suspicious account.

Can you guess the account number?
You're right. It was 666! The number when shown on the screen seemed to be 666 - and several other digits, but when the actors said the account number all I heard was 666.

This may seem totally innocent and I doubt that the writer of the show was trying to proclaim 666 as the eternal damnation number. I do think, however, that the number was purposefully used because of its association with the devil and bad. Although these two sentences sound contradictory, they are not. The first assumes that the writer was not trying to promote the Mark of the Beast as a religious idea. The second assumes that the writer used 666 simply for its mythical impact.

According to some writers and thinkers this subtle interjection is in fact a means of bringing about the Antichrist rule. The subtlety and overuse of the term would tend to break down the defenses against it. This idea seems possible. We all can remember when certain words were not allowed to be said on television and radio. When they were first allowed it was startling to hear them. Now most of us listen to a program using those once prohibited words and hardly even hear them. Some writers seem to think that this type subtlety is allowing the Antichrist philosophies to creep into all phases of our

The Beautiful, Ubiquitous Beast 107

lives including some traditionally fundamental Christian churches.

Two books I recently read attempt to alert people to the subtleties of the encroaching Beast. These books point out the way anti-Christ philosophies are creeping into our homes, church pulpits and almost all parts of our lives. One of the books describes the subtleties even to the extreme of the air freshners that hang from many car mirrors to give the car a nice fragrance. In <u>The Hidden Dangers of the Rainbow</u> by Constance Cumbey there are numerous indications of how the undermining of Christianity is happening. The book is replete with references for documenting the claims of the things that are constantly taking place today. Her points are well made and frightening. As the book indicates and as the "Scarecrow and Mrs. King" television program illustrates, we are able to see some of the things which are very subtle. These subtleties do raise questions as to the manner in which the Antichrist will take over.

In The <u>Seduction of Christianity</u>, subtitled "Spiritual Discernment In the Last Days," Dave Hunt and T.A. McMahon also make many interesting points which cause us to see a trend that, as Constance Cumbey did, alert the reader to a possible undertow of an antichrist nature that is so rapidly, forcefully, yet ever so subtlely, subverting Christianity.

But, beyond the subtlety and subversiveness of offering there is an even greater danger - IGNORANCE. How many people have never heard an explanation of the Antichrist or the Beast? Someone who has not been taught will probably think it is just another way to be taxed, but certainly nothing of eternal significance. If you are one of those that has been taught, you know that you will not have to worry if you have lived "right" and by the good graces of God you are spared this era of history. You may be one of those that has been taught, but are able to rationalize, "God just would not let it happen to me." You will probably be able to rationalize that you will be able to accept the Mark and hide it from God with makeup - remember Adam and Eve?

In most cases the mechanism for the Beast will not only be 'tres chic,' it will be convenient, money-saving and wise. There are many other, more sublime, reasons. Below is a list of reasons why you may choose to accept the Mark:

1. SHEER IGNORANCE - Know nothing of the Antichrist and don't care.

2. DISBELIEF - That it is true.

3. BASIC NEEDS - Food, shelter, clothing, and medical help.

4. ECONOMIC - No need for the extra car, shopping and banking are done at home. Simple record keeping.

5. CONVENIENCE - No keys, no money, and no checks to lug around.

6. SECURITY - Fool proof burglar systems, everyone is tracked.

7. FAD - Follow the Charismatic leader.

8. RATIONALIZATION - It is true, but God will never let it happen to me.

9. GRADUAL ACCLIMATION - It happens so gradually that you are caught up in it before you realize it.

10. RELIGIOUS FERVOR - Through emotionalism and blind faith the fervent one fails to see the world problems and the dangers of the Mark.

11. INTRIGUE - Fascinated by the technology, you forget the side effects.

12. DEMAND FOR WORLD ORDER - People have not found peace nor order in the present governments, the legal and justice systems, nor in the churches.

Can you think of a friend or relative who may be enticed to accept the Mark for one of these reasons? Is there one item on the list that could convince you to accept the Mark?
    In view of these twelve points it is so easy to see the need for the mechanism. It is also easy to see that it is not just going to happen. It is happening this very moment! Whether the plan of God or just anthropological evolution you can certainly see the Antichrist mechanism being developed. It is happening in the social world, the technological world, and the political world.
    It is easy to see that this Antichrist mechanism is also necessary for our rapidly becoming over populated world. With my knowledge of the Beast, I still agree that there should be only one identification necessary for an individual.

Now with biometric identification the Social Security Number may no longer be needed. In either case, one identification could be used for everything: birth records, Social Security retirement records, accident and health records, military service records, police records, banking services, Internal Revenue Services, mail records, and church records. In fact, it is the only identification necessary for records and information of any nature. There is no need for having a separate number for each organization or record. It makes sense for your personal convenience. It makes sense economically. And, it makes security sense. If this number were stored in one data bank of the federal or world government, each state could access it for things from the driver's records to criminal records.

In one major airline passenger computer system passenger name records are broken into six sections of a data base. These six sections are labeled A,D,H,M,Q and T. Each section has records in fixed or specified locations so that it is easier to search for a particular passenger. If a person's last name is Harris, the computer would start the search in the third fixed location, the one beginning with the letter H. As many airlines have modified these alpha groups or data base sections to five, four or otherwise, so could the government.

Using computers the airline industry and government have mastered data base and information center concepts. Even without a computer, you have made use of the concept yourself. The telephone directory, a recipe list, a greeting card list or any other list basically uses these concepts.

An encyclopedia is a very good example. Depending upon the encyclopedia and the publishing company you may have a set broken down as follows:

or this same information could be on single pages in file cabinets:

The Beautiful, Ubiquitous Beast 111

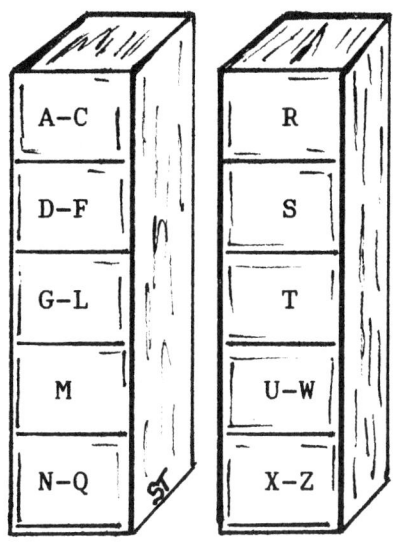

or the encyclopedic information could be stored in files on computer disk:

In fact, it could be set up so that one computer is used for each letter of the alphabet. There may be one computer that searches for the one name starting with A, another for the name starting with B and so on. The name lists can also be set up so that it may be accessed either by name or social security number. Lists are easily maintained.

All types of data can be kept about you and me.

Maybe we should say, "Thank God for government beauracracy." The reason: the United States government has some of the sloppiest computer organizations in the country. In my opinion private industry runs circles around the government without trying. If the government could use a few of the good managers from industry for implementation of the latest technology in computers and communications the Big Brother of George Orwell's <u>Nineteen Eighty Four</u> would be with us. If you have read <u>Nineteen Eighty Four</u> or have seen the British movie based on the book, you may ask, "What is the difference between Big Brother and The Beast?"

The answer is, "NONE."

ORWELL, BIBLE WRITERS AND THE LIST

Orwell, like Bible writers, was writing without "hands-on" experience of the mechanism of the Beast. Even as I write, things of technology are being updated. Computer and communications technology is being updated so fast that it presents a dilemma for managers in the business world: "I need a computer today, but if I buy one today it may be obsolete tomorrow."

Thinking of item eleven in the list on page 109, Intrigue, the computer certainly may be a part of the subtlety which will bring a person to use the Antichrist mechanism. But, beyond mere intrigue, the benefits of the computer are so easy to see. These benefits will aid in selling the MARK to most people whether they are educated or ignorant, rich or poor. When I tell you that the computer will be the thing that relieves the United States of its dependence upon foreign oil, you may think I'm crazy, but when you read the explanation later, I think that you will

The Beautiful, Ubiquitous Beast

agree. You may not have heard, but the computer will virtually eliminate the newspaper as we know it today. The computer is changing our way of life and it is intriguing. When you have finished this book you will probably be afraid of computers, but you will have the insatiable urge to buy one for yourself!

Before you buy your little home computer, ask yourself, "Am I helping the Beast?"

Yes. You are! But, if these stories of the Beast are true, it is going to happen whether you like it or not. "Que sera, sera, Whatever will be, will be...." Remember the song?

As the song states, so some fundamental Christians believe. It is God's will and it will be. And, there will be benefits!

Many of those benefits we use each day without even thinking about them. One benefit that will affect everyone comes under the name of Point of Sale. The term means exactly what it says. It is the place where you pay for what you buy. It is the point where the buyer pays and the seller receives. The Point of Sale in your grocery store is the cash register.

You may wonder, "Why make such a big deal about something so insignificant as the Point of Sale?"

Well, think again!

The Point of Sale is the most important part of any business scene, but how about in view of the awakening Beast. Considering the awakening Beast, Point of Sale is one of the main points of control. Although I referred to it as the cash register spot, eventually there will be no cash register. There will be no cash. At first, the cash register and a small debit/credit card reader will be on the checkout counter together. Then the cash register will disappear. At that time there will be no money exchange. Using Universal Product Codes, as discussed earlier, and card readers, the check out person will need very few skills. The technology eliminates the need for skills and makes everything so convenient for us. From the list on page 108, items four, five and six were: Economic, Convenience, and Security. Each of these is a good reason for a Point of Sale system and a very important piece of the Antichrist mechanism puzzle. The benefits of computers to the airline industry were incredible. The benefits are just as incredible to your bank and to your grocery store. The same type system used by the airlines is used for some point of sale and bank teller machine systems. In addition to the need for fast messages like those in the airline reservations system, there is also a need for extreme accuracy when dealing with money. This has been taken into account by those developing Point of Sale and another computer system, Electronic Funds Transfer (EFT).

EFT is the electronic transfer of money. Usually when we think of money we think of paper bills or metal coins. If we have never had the occasion to deal or trade in places other than the United States we would probably only think of dollars and cents. If you were in Russia you would think of roubles and kopecks.

But, what is money?

Money is anything that is used as a medium of exchange. Salt, shells and various other things have been used throughout the history of the human race. Today, most of the world uses some kind of paper money which is regulated by the government of that society. Whatever the medium of exchange, it must have a standard of value. If we are using shells as that medium we must know the worth of one shell. For example if twelve ounces of Pepsi Cola costs three shells, a person knows that he must pay four shells for sixteen ounces of Pepsi Cola unless the local grocery store is running a special on sixteen ounce Pepsis. There is a certain value placed on each of the shells. From the value given here in the purchase of a Pepsi we readily see that it

The Beautiful, Ubiquitous Beast 115

would take a truckload of shells to buy a car.

If you are paid with shells each week, you probably do not receive enough to buy a car. You save till you do have enough. Therefore your shells or your money is a way to store your wealth. You are able to go to the safe and look at your dollar bills or your stack of shells. In years past many people that were able to save any money had it stashed in jars. They could look at their wealth every day.

THEN CAME BANKS!

A person could no longer actually see his money every day, but he could look at his bank balance and know what he had. Today money is becoming less and less visible to the owner. Before it is totally invisible, it will be represented by a plastic card.

"Plastic Money" is a term which people often use when they speak of a "moneyless society." Although the society will not be truly moneyless, money will be stored as invisible electronic bits in a computer. When you normally would have deposited money to your account, it will now be the addition of electronic bits. You will be able to buy against those bits through the use of your credit and debit cards or your 666 - biometric identification.

Does it sound BEASTLY? It is BEASTLY! It is happening because of our intrigue with computers and our need for security.

SECURITY?

Yes! One main reason for the use of credit cards, checks, and other non-currency forms of money is safety. It is safer in our society to have no money. On July 9, 1986 there was an article in the Gaffney Ledger stating that one of every four households in the United States was touched by crime in 1985.

THE WAYS AND MEANS

Everything is working toward the mechanism necessary for the Antichrist state. Technology and human ingenuity have given us the means to accomplish the Antichrist state in a moneyless society; and, even crime has given us a reason.

The moneyless society became more realistic when the banks, credit unions and employers began to offer the special service of depositing your payroll check into your bank or credit union account. This eliminated your need to go to the bank or credit union every payday.

The SUBLETY OF THE BEAST is witnessed again!

This new benefit is great for the employer, the bank and YOU! It allows you, the employee, to use the banking time for pleasure or other duties of your life. To the employer it saves time and money in bookkeeping. To the bank it means more business with a less expensive process. It also eliminates the need for you to drive to the bank. Do you remember an earlier statement that the computer would be the way in which the United States ends its dependence on foreign oil?

This automatic deposit idea reminds us of what we have heard about the Great Depression and the stock market crash in October, 1929. People did not trust the banks with their money. They kept their money in fruit jars or other places of "greater security". Only a half century ago many of the large factories paid their employees with envelopes stuffed with cash. Before that period we recall the "good old days." There were stagecoaches and trains that delivered the payroll in gold.

THINGS HAVE CHANGED!

Things are still changing, but many people argue that the Beast situation will never exist. They argue this because the prophetic stories and sermons indicate that the Beast situation should be worldwide. They argue that even world language barriers would prevent a BEASTLY society.

Would it?

The Beautiful, Ubiquitous Beast 117

The language Esperanto, invented in 1887, was an attempt at a universal language. Although it may not be considered a raving success, it is an improvement over some other attempts and an indication of man's desire to have a world language.

It is also important to remember that most languages of today did not arrive at their present state by plan. With many variations, English has almost become a universal language. Pilots and air traffic controllers use it around the world. Even in places like India where Hindi was once proclaimed the state language, English is still the de facto business language.

In the United Nations there are six official languages: English, French, Spanish, Chinese, Russian and Arabic. When a speech is made in the General Assembly these languages are the only languages acceptable. When the speech is made in one of these languages instant translations are made into the other five. The delegates to the General Assembly wear head phones to listen to the language of their choice. In the International Court of Justice, English and French are the two languages acceptable for presenting speeches.

Were these languages chosen for political reasons?

No! Not entirely. They were chosen because almost every world citizen of any literacy reads one of the six. Almost in a Darwinian fashion French and English have become the commercial languages of the world. It is also important to remember that there is a language that has evolved over the centuries that is almost universal.

That language is the language of mathematics, especially the arabic numbering characters: 1,2,3,4,5,6,7,8,9, and 0. In many parts of the world one million dollars would be written $1.000.000,00 and in other places it is written $1,000,000.00. The periods and commas are used differently in separating the dollars and cents. Although these differences exist they are much easier to overcome than those such as Ohayogozaimasu, Buenas Dias, Bon Jour, and Guten Morgen which mean Good Day or Good Morning in Japanese, Spanish, French, and German respectively.

As the world becomes smaller through better communications such as television and satellite systems, language differences will continuously diminish.

A good example of this is illustrated by my encounter with a man from Glasgow, Scotland. As we talked in a noisy club on a little Greek Island, I could hardly understand him. When we walked outside I realized that it was not the noise. It was his accent! Somewhat embarrassed, I asked him if he had the same difficulty understanding me.

He said, "No!"

He explained: "I hear all United States accents. Almost all movies are made in Hollywood. These Hollywood movies have a good cross section of the United States accents. The United States people seldom hear the Glasgow accent which is worse than the New York, New York accent with its lack of grammar and enunciation correctness."

The fact that I was talking to the guy from Glasgow was in a very small way beginning to diminish these differences. The further these differences diminish the more realistic it will make international monetary exchange and an international language. Every world citizen will use such things as Point of Sale.

Item number five of the list of things that may convince us to accept the Mark was Convenience. As mentioned earlier Point of Sale is one of those conveniences.

Point of Sale is a concept born of a need for a commercial leading edge. When first conceived it was probably the idea that one bank in one area would be involved. That concept has expanded tremendously. The essence of Point of Sale is that neither the customer nor the business has to be bothered with money

The Beautiful, Ubiquitous Beast 119

and banking. As a natural, one of the first businesses to experiment with Point of Sale was the grocery store. As you have probably heard, some of the persons inbred with the fear of the Beast are awed to cold chills when they go into one of the grocery stores with the optical character readers connected to the cash register. The clerk drags the items across the reader and the computer cash register does the rest.

Taking this further, Point of Sale systems will diminish language barriers. There will be no problem in converting your dollars into German Marks or other currencies because the Point of Sale computer system will be programmed to make the conversions automatically. Some credit and charge card companies already do this. You may buy an item in Germany, priced in German Marks, but when you get your statement in the United States it is converted to dollars. This will become very common as long as differences in money exist. The Antichrist will eventually have one universal, invisible money. His money will be the world money!

UNIVERSAL, WORLD MONEY!

There is one! It is the SDR, of the International Monetary Fund. The SDR, Special Drawing Rights, is an existing international form of money. The value of an SDR is based on a pool of currencies such as the dollar, the French franc, the deutsche Mark, the pound sterling and the Japanese yen.

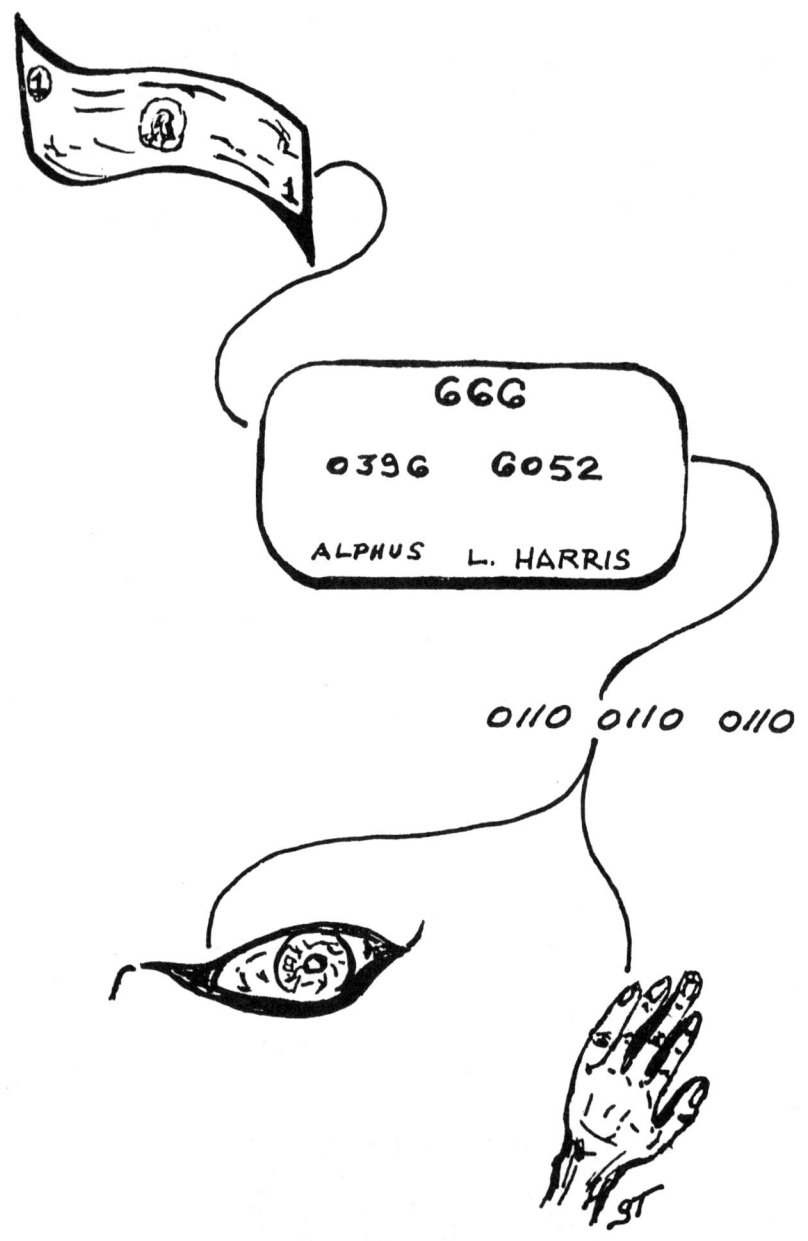

CHAPTER 10
WORLDWIDE SHOPPING

To exemlify Point of Sale let's pretend. Take out your Point of Sale card. It is technically called a debit card because your account is debited. When you make a purchase and use the card, money is taken from your account and applied to the store account. When you finish your shopping and go to the cashier, he drags each item across the glass eye in the counter. The price of each item is added and the computer voice tells you the amount of your purchases. You hand your card to the cashier who pulls it through the slot in the telephone gadget.

The cashier looks up, hands you your card and says, "Thank you," as the un-informed person standing beside you wonders why you are allowed to take your groceries and never pay for them.

The things that took place are described below:

- Your card was put through the telephone gadget.
- The card activated the phone line connected to the bank computer.
- The bank computer read your grocery bill and identification number.

- The bank computer checked your balance.

```
┌─────────────────┐   ┌─────────────────┐
│ Your            │   │ Grocery Bill    │
│ Account         │   │                 │
│                 │   │                 │
│                 │   │                 │
│ Total   500.00  │   │ Total    35.00  │
│ ><><><><><><><  │   │ ><><><><><><><  │
└─────────────────┘   └─────────────────┘
```

- The bank computer transferred the money from your account to the store account.

```
┌─────────────────┐   ┌─────────────────┐
│ Your            │   │ Store           │
│ Account         │   │ Account         │
│                 │   │                 │
│                 │   │                 │
│ Total   465.00  │   │ Total    35.00  │
│ <><><><><><><   │   │ <><><><><><>    │
└─────────────────┘   └─────────────────┘
```

- The computer told the cashier to let you have your groceries.

This little scenario does not seem ominous nor BEASTly. Yet, let's expand our example from this local situation to a national or international banking situation. Like the airline computer systems, the banking computer technology is intricate. It affects more people and demands more sensitive and personal data. And, it certainly affects the one thing for which most of us are concerned - MONEY!

In the Electronic Funds Transfer system, one large bank may act as the clearing house. In seconds, billions of dollars can be

Worldwide Shopping 123

transferred around the world. In the same manner information about the customer can be sent from California through a clearing house in New York to a bank in New York, Columbia, S.C. or Lusaka, Zambia.

Even when you consider the whole world, Point of Sale does not seem too ominous. The Big Brother or BEASTly part is not obvious until you consider the information that must be kept for each of us. Then when you think of the control of each of us that the information gives the BEAST, it is ominous.

In building an information base on all of us, the corporation or government will use many computer and communications concepts. Some of the concepts such as Electronic Mail, Office Automation, Information Centers and the massive networks of communications have an astounding impact upon our lives.

Consider the frustrations caused by the United States postal system. Have your letters ever been six months late? Have you ever expected mail and never got it? Have your letters first gone to someone in another state? These problems will cease with Electronic Mail. We will love it! To date it is mostly used by businesses, but personal use is not far away.

Electronic mail is basically the ability to send a letter or message from one computer terminal to another. In home use this is probably a letter typed on a personal computer and sent by modems and telephone lines to a friend who also has a personal computer. Somewhere between you and your friend there is probably a big computer where your letters are actually stored. In familiar terms: your personal computer screen is your stationery and local mailbox. You type the letter and then depress a button. This puts the letter into storage in the big computer just as a mail person normally takes the mail from your box and puts it on a truck, train or airplane to the main post office.

At the main post office the letter is

routed to the mailbox of your friend. Electronic Mail is similar. Your letter gets to the large computer, and is put into your friend's mailbox. Using his little computer, your friend can read your letter any time, just seconds after you have written it.

As in the airline and banking industries, the main computer may be anywhere in the world, but with modems, telephone wires and satellites it is as if it were next door. With "registered" or "certified" designations you automatically receive messages telling you exactly when your friend opened his computer mail box and read the letter. You know that your letter will not be lost. Electronic Mail will make correspondence a pleasure.

With Electronic Mail, Office Automation (OA) is becoming a part of the corporate electronic system concepts. OA, in simple terms, is getting rid of paper work, thus the term, "paperless office." It is an effort to streamline office work using electronic equipment. Letters are typed using word processing computers. Files are kept in electronic filing systems. Electronic Mail is used for communicating. Proposals are dictated into machines for the wordprocessor operators to type and mail. The list is so vast that there are magazines dedicated to Office Automation.

Electronic Mail and Office Automation fit snugly into the BEAST mechanism. So does the Information Center concept. When in the banking atmosphere I thought the Information Center potential for my company was fantastic. The one who posseses proper information posseses the power necessary for financial and political control.

This control may start with information from Charts of Accounts. The data from these lists of customer accounts can be used by credit managers to determine whether to lend money to a client. The decision will be made on that local data. However, with an Information Center, the credit manager could

Worldwide Shopping   125

have access to the client's financial statements in all regions and a better financial picture of the whole client company. The information may change the credit worthiness of the company tremendously.

FORGET BIG BUSINESS! Think of your own account. If you live in Blacksburg, S.C., and have an account at the local bank, Harris Auto Parts, Inc. will probably accept your check for your purchases. However, if you travel to Springdale, Arkansas it may be hard to cash that Blacksburg check. With Information Centers and Point of Sale systems you will be able to use the money in your bank account whether you are in Blacksburg, Springdale, or Kalispel, Montana.

If Harris Auto Parts, Inc. has the Point of Sale terminal, the telephone gadget for reading your debit card, you no longer need the check to buy the goods there and Harris Auto Parts, Inc. no longer takes the risk of having your check bounce. The funds are transferred immediately from your account to the account of Harris Auto Parts, Inc.

Now, get into your car with your Point of Sale card and travel to Springdale, Arkansas. Walk into Debra's Auto Parts Store which is equipped with Point of Sale terminals. When the clerk and optical scanner system reads and counts your purchases the clerk drags the card through the telephone device, and in nanoseconds (one nanosecond = one millionth of a second) the clerk hands you your card and your purchases. The Point of Sale information query left Springdale, went to Little Rock, then to New York, down to Columbia, S.C., and into your bank in Blacksburg. Fortunately for you, your account was great enough to cover your Springdale purchases. The verification traveled the reverse path to Springdale!

Does the HEAD OF THE BEAST appear closer?

You may say, "Yes," but for many, "We are just using our brains and the resulting

technology of our brain power."
   Many will not believe that he is Antichrist and others doubt that such a creature would ever be accepted.

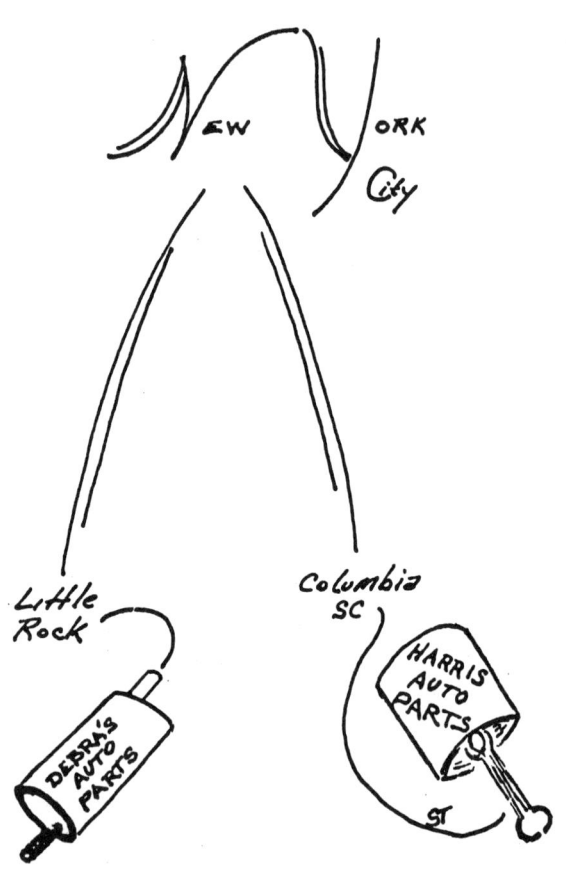

CHAPTER 11
WORLD CITIZENS ACCEPT THE ANTICHRIST

He will be accepted!
The acceptance and unquestioned reign will last almost three and one half years. Then the Antichrist will begin to show his true evil nature and the Jews will realize that he is a fake. He is an imposter of the Messiah, maybe an imposter Jew. Recognizing their mistake of two thousand years the Jews will accept Jesus Christ as their Messiah. This is why many feel that during the Antichrist reign there will be very few with the fortitude to withstand the pressure to accept the MARK. Others think that the only ones who will refuse the Mark are those that are hearing about Christ for the first time.

Most Gentiles will rationalize how they can take it and still be able to make it through the pearly gates. For the fundamental pentacostals it will be pure rationalization, but for the Roman Catholic it may be easier to understand because he expects to sit in confessions and wipe out the prints of the BEAST. The Southern Baptist will be able to quote St. John 10:28 to reason that no man shall "...pluck them out of my hand" of eternal Grace. Whatever the rationale, many will fall by the wayside as the Jew again stands in triumph.

Remember, the Jews are "God's chosen people!"

Christians may believe this, but Jews do not always agree with Christians on their thoughts about Jews. Didn't Golda Meir once quip, "If we are God's chosen people, why did he lead Moses into the only part of the Middle East which has no oil?"

Another reason that it may be easy to dupe people into recieving the MARK is that churches may eventually require the identification MARK for entrance. Who could believe that his own church could be a part of the forces of the Beast. Churches will accept the BEAST mechanism. Think about your place of worship and crime. Do you get the picture?

Most of the discussion of the Beast mechanism has been quite obviously in the interest of bettering our quality of life - saving time, money, banking headaches, et cetera.

But, for the churches the Beast mechanism will be a deterrent to vandalism. It will become a necessity. The churches, too, will join the forces of the Antichrist. Without the help of the churches the Antichrist could not take over.

If you think this is only the belief of the fanatics, count the fanatics you know. Fanaticism is one trait that seems to transcend all religions. The strength of any organization probably comes from its fanatics. The lukewarm members would allow the organization to die. Even the Bible, in Revelation 3:16, depicts God as despising the lukewarm and "...will spue thee out of my mouth...."

An illustration of the kind of fanaticism that seems to exist in all religions follows:

Once at the U.S.- Arab Chamber of Commerce convention in New York, I was in a group academically discussing philosophy and religion. I said, "When Mohammed wrote the Koran...."

Immediately one young Moslem man verbally pounced upon me, "Mohammed did not write the Koran. He transcribed it. God wrote it."

Fortunately, an Iranian, Moslem friend of mine was in the group and knew that I was not demeaning the holiness of the Koran nor Mohammed. He intervened in my defense and told the young man that he should learn the English language and that I meant nothing derogatory.

This devoutness or ignorance exists in all

World Citizens Accept The Antichrist 129

religions. Watch the people at a Catholic shrine such as Fatima in Portugal. As mentioned earlier, some of the devout worshippers walk around the entire building on their knees and light candles. Almost to the other extreme in worship are the Pentacostals praying loudly and fervently asking God for healing or soul salvation.

   The sincerity is great, but many of these sincere people have a strong blind faith. They will follow a minister, priest or Rabbi regardless of how riduculous and foolish he is. In the ninth chapter of St. John we see that it is better to be physically blind than to be spiritually blind, yet do you know more spiritually blind or more physically blind people? Many spiritually blind are also politically blind. They will follow a politician and vote time and time again for a person who has pretty well proven that he has no interest in the voter except his vote.

   People follow politicians, preachers, rebel rousers or anyone who will step to the front of the crowd. Once the fervor is aroused the followers blindly let that fervor flow into all parts of their lives. To the fundamentally religious, that is how it should be. But, as with the young Moslem man, at times the fervor is so hot that it melts all common sense.

   An incident in my company cafeteria further emphasizes the blind fervor. One day, I sat with a black friend. Everyone at the table, except me, was black. A personnel recruiter for the company, said that he was searching for blacks as part of the company's equal employment opportunity drive. He asked if I knew any blacks that would like jobs in the computer department. Unfortunately, I responded!

   "Yes, I know a boy in Shelby, North Carolina. He asked me about our computer department. I'll give you his name and number if you would like or I will have him call you."

When I finished, everyone at the table was glaring at me.

My friend said, "Oh, come on you guys. He didn't mean anything!"

I sat there for several minutes trying to decide how I had offended everyone. I tried to remember if Shelby had a special significance in history as a bigot and unfairly discriminatory area. I could not remember anything.

When we left the table I asked my friend to explain.

It was simple! I had used the word, "BOY," which to many blacks is bad. Although "boy" still has no negative meaning to me, I do not use it when referring to black males.

The past few paragraphs illustrate how blind, fervent people will follow anyone who promises a care-free life style or has charisma to arouse emotions. There are many of those leaders today!

Jerry Falwell has many followers nationwide who do more than just listen to sermons. They send money when he asks for it. According to an article on page B3 of the Washington Post, 12/2/84, which referred to court papers filed in Falwell's $45,000,000 case against Hustler magazine and Hustler's owner Larry Flynt, Falwell used copies of a Hustler magazine parody of himself to raise $800,000.

Several months ago the Washington Post reminded us of the anniversary of the mass suicide spectacle in Guyana. We were reminded that California Congressman Leo J. Ryan was shot and killed in Guyana by followers of Jim Jones, the leader of the Peoples Temple. Following the ambush of Ryan and his party, Jones led 913 of his followers in a mass suicide.

Shocking almost everyone just two years after this incident, a daughter of Congressman Ryan became a follower of Indian guru Bhagwan Shree Rajneesh. It is not easy to see how

World Citizens Accept The Antichrist 131

people follow a man like Jim Jones, but they do. People who claimed to be Christians followed him even though he had admonished them to listen to him, not to the Word of God. People listened. They listened, and followed him, even to death. Do you doubt that people who have heard about the Antichrist all of their lives will still bow to him?

We say that Jim Jones and his people were a very small part of this country. We try to downplay the impact of such a little group, but, wait! Let's jog our memories.

Over the Thanksgiving holidays in 1984 I was in San Diego. Two friends of mine were going to a Terri Cole-Whitaker meeting. I was told that her Sunday morning congregations are 5,000 people or more. In addition to her local congregation she had a nationwide audience through the air waves. Allegedly her sermons teach that we create our own environments.

A friend, in a state of mockery, laughed and said, "I made the lightning strike the tree so it would fall on me."

I have listened to some of her television broadcasts. She talked a lot, but seemed to say nothing. Yet, thousands follow her. (Note: I understand that she has now decided not to continue preaching.)

When A.C. Bhaktivedanti Swami Prabhupada, leader of the Western Krishna movement, went to Toronto, it was reported that:

> Followers threw themselves to
> the floor to kiss the feet of
> the small, flower-wreathed,
> dhoti-clad figure.... The
> founder of the International
> Society for Krishna Concious-
> ness was greeted by the clash-
> ing of cymbals, the beating of
> hand drums, the bugle-like
> shriek of conch shells, and
> the waving of peacock and
> yak-tail fans.

According to their own press releases, the Divine Light Mission numbers ten million. Their leader Guru Maharaj Ji proclaims himself the "Christ Incarnate." (reference: John W. White)

Maharishi Mahesh Yogi claims that his Transcendental Meditation (TM) movement is not a religion. Allegedly, however, Yogi privately claims to be an "incarnate christ" who came from and is returning to God. (reference: John W. White)

Another group leader appearing in recent years is Sun Myung Moon. He is from Korea and allegedly has 20,000,000 followers convinced that he is the Messiah. His Unification Church in New York City reportedly can draw a crowd of 50,000. In very recent months this leader met with very strange circumstances. The government tried to disallow him religious tax exempt status on much of his property. He said that it was just an attack on him and his religion. Though this may be likely, the unlikely picture is the conglomerate of supporters he had in the matter. It included the fundamentals, the high churches and other citizen groups.

There is the Rastafarian sect led by a "black messiah."

And then there is Charles Manson! Yes, the Charles Manson of the horrible Tate murders. In the first part of 1986 Manson was up for parole. There was an outcry of the California citizens and he was not allowed to go free.

Why do I mention him here?

Although there seem to be few disciples of Manson, there is Lynn (Squeaky) Fromme. She reportedly attempted to assassinate President Gerald Ford. She stated, "We're nuns now, and we wear red robes. We're waiting for our Lord, and there's only one thing to do before he comes off the cross, and that's clean up the earth. Our red robes are an example of the blood of the sacrifice." The Lord of whom she spoke is Charles Manson. The cross is the jail.

EST (Erhard Seminar Training) was founded by Werner Erhard, formerly John Paul Rosenberg. In the past ten to fifteen years EST has become a multimillion dollar organization. From friends and acquaintances that have paid hundreds of dollars for the "screaming" week ends, I understand that the seminars basically teach that you have to face the world with all you have. Don't be concerned about others. It's you yourself for whom you must be concerned. It seems to be the antithesis of the Golden Rule. Instead of "do unto others as you would have them do unto you" it is "do it for yourself or to them first."

These are only a few of the leaders who have flocks following them, but the list is endless. Some of these leaders proclaim to be messiahs. Some claim to be businessmen using psychological techniques. Some are murderers. Some are fundamental preachers. Some are big time electronic church leaders. Several seem to have the Antichrist spirit, but are too gross to be the Antichrist. However beautiful or ugly, it is obvious that a great many of our world citizens will bow down to anything.

CHAPTER 12
ANTICHRIST SOCIAL CLIMATE

It seems that the social climate is right for the Antichrist; people will accept him; the political setting is right; the economic picture is right; the world religious stance is right; and, society is prepared to accept the Antichrist! Are you?

We have discussed murderers, preachers, and others whom people follow. We have discussed politics, religion, and technology. In fact, we have discussed enough to make the Antichrist reign seem very feasible. To many we have probably discussed enough to end the book and start preparing for the inevitable, the Antichrist. Yet, there is more! It is awesome to realize how many are ready for the Antichrist or as the believer may say, "Unprepared for the return of Christ, the real Messiah." When you think, "Every 'important person' is ready," you may shudder and say, "Never me!"

But, are you? Just who is ready?
Are the:
- Financial wizards of the land?
- National government officials?
- State government officials?
- Local government officials?
- Foreign government officials?
- Scientists?
- Educators:
    - College?
    - High School?
    - Grade School?
    - Kindergarten?
- Clergy and/or Churches:
    - Roman Catholic?
    - High Protestant?
    - Fundamental Protestant?
    - Jewish?
    - Moslem?

Hindu?
Lawmakers?
Law Enforcement Officials?
Transportation Officials?
Housing Authorities?
Environmentalists?
Prolife groups?
Newspaper people?
Philosophers?
The General Public?

YOU?

The next few pages illustrate why the time of the fundamental Christian predictions of the Antichrist are here.

SEX SURVEY

Recently a survey by columnist Ann Landers showed that 72 percent of the women who responded to her survey request preferred cuddling and hugging to sex. Forty percent of that 72 percent were under forty years of age.
    You may ask: "What does that have to do with the Antichrist?"
    It is the beginning of an explanation of how the world today is as it was in the days of Lot, and in the days of Noah. The world is very much like the twin cities of Sodom and Gomorrah. It is not the sexual aberrations of which I write. Sexual craze is merely a manifestation of the true problem.
    The survey results seemed to surprise the columnist. It indicated that even in "normal" sexual relations between partners there was little feeling between the two. Sex was more an animalistic act than an act of love. In the days of Lot, people were totally unconcerned about the feelings or needs of their fellowman. That is the root of the problem and the major reason that the Antichrist is so near.

Antichrist Social Climate 137

    The columnist expressed surprise. The results of her survey were a sad statement on relationships between couples today. I believe her evaluation is right, but it may be just as surprising to do a survey of men. It would not surprise me to know that men responded similarly.

    There seems to be feelings of inadequacy among men. There is tremendous pressure to perform. It seems that the actual performance of a sexual act has become the most dominant mental picture of a relationship. Love is not sex! Sex is not love! Sex could enhance a love affair and create more solid partnerships. Yet, sex is usually only mentioned in smutty talk or condemnation.

    Sex is blamed for the falling away from the family unit. But, the family is the reason for that. So many "Christian" families display such unhappy, uncomfortable and un-Christian role models that youngsters see no need for such a lifestyle. We are developing into a cold, inconsiderate society.

    In Sodom and Gommorrah there was no concern for the neighbor, totally contrary to Bible teachings that you should "...love thy neighbor as thyself...." (Leviticus 19:18). To love someone requires that you respect the rights and property of that person. Sodom and Gommorrah citizens did not love their neighbor, Lot. They did not care for anyone. They were interested in their own immediate pleasures. In Genesis 19:9 they did not even respect the rights of the angels of God who had come to visit Lot.

    Many people today are like that. We even have a nice word for that attitude. It is hedonism which means regarding pleasure as the highest good, or devotion to pleasure as a way of life.

    There is no wrong in pleasure. It is just that pleasure really has no definition. Each person defines his own pleasures. Today and in Lot's day, that definition seems to evolve for each person without respect to the needs of others. It appears that we have developed a society of egocentric people with the motto, "Give me pleasure. Be damned world."

In his autobiography, Chrysler Corporation Chairman, Lee Iaccoca mentioned that the reason for many of our problems is greed. He referred to greed as the worst of the seven deadly sins.

In an interview with a leading U.S. magazine, Japanese billionaire industrialist, T. C. Wang, spoke of decadence of America and he was not implying sex. He was speaking of decadence in our feeling of self pride, ambition, goals and our general lack of purpose. In that way the United States is like Sodom and Gomorrah before the final fire of Genesis 19:24.

Earlier I stated that many will ask for the Beast. Above I have listed many groups that appear to be ready for the Beast. Not only are they ready, but they are crying for the Beast. No they are not saying the words, "Come Beast." But, that is the plea of their actions. Even churches are changing to accommodate the Beast.

Not long ago there was an article in the Wall Street Journal about the Electronic Church (E- Church) and the many preachers (E- Ministers) that appear on television and receive tremendous amounts of money from viewers. The E- church was discussed from a financial standpoint as is typical of the Journal column. Yet, I feel that the Journal discussion of the E- church was hardly of the one we will see in the near future. Radio and television preachers should study the next few sentences. They may find just the thing for continuing the good work they are doing - at least building the bank account.

Please E- Ministers, forgive my cynicism. Probably few of the people reading this would ever send you money anyway. Instead of hating me, you may want to contact me as your consultant to get your fantastic electronic networks in place to insure promptness of your

Antichrist Social Climate

suppporters. THERE IS LITTLE TIME BEFORE THE BEAST WILL TAKE AWAY ALL MONEY MAKING POWERS.

As described in the Wall Street Journal, the E- Church is the minister coming into the home via television. Not described in the Journal was the way that listeners would be able to respond to the E- Minister immediately. Soon you can respond "NOW" to his plea for money!

If you have not seen the computers that use the television screen as the video display, you may be surprised to know that television and computers have been attached for games and home business needs. One minute you can be watching your favorite program on television and the next typing away at a keyboard as the typed messages and the calculations appear on the same television screen.

Now think!

How will this enhance the E- Church?

Turn on your television Sunday. Listen to the inspiring sermon. You are moved to give, to help this worthy preacher, especially if he has told you that God is going to take him away from you if he does not raise $8 million in the next three and one half months!

You immediately go to your keyboard (typewriter like gadget hooked to your computer and television) and check your bank account. Yes, you have enough to send to this worthy cause. Fortunately for that preacher you have acquired all the amenities of the rearing BEAST. You decided months ago to opt for home banking and you are able to transfer your funds immediately into the account of the preacher. On the screen, just below his mailing address, he has flashed the computer code for you to deposit directly to his account. That is not the end of the miraculous electronic wizard nor the suave of the electronic preacher. You immediately receive a message of thanks for your contribution on your screen. You see the concern of the preacher because he responded immediately.

But, wait!

We are discussing the Beast and his cunning ways. The preacher may have been basking in the South Pacific sun for three weeks and his offices closed.

Then who responded? The Beast?

Yes, you guessed it. Remember the airline and Point of Sale systems. The computers "talked" among themselves without human intervention. Well, we may call this "computer ministry" because the computers are talking again. As soon as the one controlling the ministers account realizes that a contribution has been made it responds to the contributor. Yes the computer, the hand of the Beast, responds to you and tells you that God will bless you for your generosity. And, he will pray for you while the preacher basks in the sun. The sermons come to you from a library of video recorded sermons and a computer sends you the "blessings of God." Maybe the computer is programmed to pray for you!

Now, if that has stirred the ire in you, look at other de facto benefits of the computer. You now are able to correspond directly with the preacher through the network onto which your home computer is connected. Yes, through Electronic Mail you will be able to have the message, prayer request, or counseling needs in the preacher's electronic mailbox within seconds after you have had the urge to seek his help. There will be no need to fear a lost letter as you may in the U.S. mail. With Electronic Mail you can trace your letter to the electronic preacher and know exactly when your message is read. But, when can you expect the reply?

Immediately! Your reply will be instantaneous, but, who replies?

We said the preacher, but that may not be the case exactly. Someone else with the proper identification codes for reading his mail may

Antichrist Social Climate

be the one that has answered.

Or....?

You guessed it. The computer analyzed your letter for verbs and nouns and selected a response from a library. The preacher may still be in the South Pacific. Currently some of this is still not perfect. The computer chooses a word that no human would ever choose. Even though these mistakes occur, there are millions of these letters that go out correctly. In the near future a field of study called Artificial Intelligence will eliminate those computer "human" errors. In the early 1980's the Japanese budgeted billions for research in this area. Almost all American companies have an interest in this area. In the Time Line chart on pages 72 through 75, we see that Texas Instruments announced the first Artificial Intelligence on a single chip. As Artificial Intelligence technology gets better and better the errors will be found less and less. Remember that only a few years ago we didn't have a computer to write letters!

To put this into perspective of our everyday lives, consider some of your daily mail. It may be like that of Johnny Carson. As I recall, one night Johnny read a letter to his company, Johnny Carson, Inc. The letter began very warmly with, "Dear Mr. Inc." Later in the letter it further personalized it by inserting Mr. Inc. into the body of the letter.

You have probably received similar mail. A few days ago one of the big 'you have won sweepstakes' arrived at my address. Mr. Harris A. Parts had won. Why did I receive it? Because I was the president of Harris Auto Parts, Inc. The computer had personalized it by considering Harris as the first name and initializing the second name Auto. With further perfection of Artificial Intelligence this type error will cease.

Yes, we see it coming! It is getting more perfect with each new concept and invention. We discussed Artificial Intelligence, Electronic Mail, Social Security Numbers, Electronic Churches, Point of Sale, Electronic Funds Transfer, Banks, Airlines, Biometrics and other ideas and technologies. We have discussed social situations which seem to foment the Antichrist mechanism. Yet, there is one simple piece of equipment which gradually draws us all closer to the Antichrist! That little piece of equipment is likely to be in your own home now, on the table or on the wall.

Yes practically everyone uses it - the telephone!

CHAPTER 13
PHONEMES
HOLOGRAMS
DIRTY POLITICS

Yes, that little telephone!

In 1876 Alexander Graham Bell had no idea that his little invention would ever be a part of anything BEASTly, yet the telephone and computer seem to be the center of the BEAST mechanism.
 Have you ever called a telephone number that is no longer in service? If not, try it! Listen to the recorded messages. The messages are recorded, but when the number you dialed is inserted it is usually a synthesized voice. It is not a human voice at all. If you are aware of this, you recognize them immediately. The synthetic voice quality is being improved daily. Soon you will not be able to distinguish it from a human voice.
 Already there are recording devices for storing phonemes. Phonemes, pronounced fone - ems, are little pieces or bits of sound. Phoneme recording devices are extremely sensitive. When they are perfected you will be able to take the phoneme library of your favorite, but deceased, singer and have him or her sing a new song for you. Yes your dead, favorite singer may perform for you again. In fact, when you write your own song you may want your favorite singer to sing it. He will be able to, dead or alive.
 Do you see the Beast raising his head?
 Can you imagine that one day you will hook your home computer and video monitor to the international communications network to hear your favorite preacher, dead or alive, preaching any sermon at any time. At the same

time he may be telling someone else something different. In fact, as these advances are made, that preacher will thank you vocally and personally from the television screen. He can even call you by name.

In addition to phonemes and voice synthesizing, graphics computer software will allow you to see your favorite preacher animated, an animation indiscernable from a television live appearance. Hologram technology will add more real-ness to the missing preacher. You may sit in an auditorium and listen, and watch the speaker, but, DON'T TOUCH! The person is not there!

With perfected hologram and phoneme technologies a politician will be able to make different speeches simultaneously over all networks. He can say to each section of the world exactly what that section wants to hear. All sections of the world may hear one person saying different things, in different languages at the same time!

Frightening? YES! It is DIRTY POLITICS PERFECTED!

Whether one believes in the Antichrist or in God, an Antichrist social climate is rapidly approaching. Just twenty years ago most of this was science fiction. Today it is reality. The fancy technology which we use daily will help the Antichrist to be all that he can be. It will make him suave and very appealing to the people.

PEOPLE LIKE TO BE LIKED! A person likes to feel that he or she is a part of the group. Teen-agers strive to be like their peers. Each year almost a half million youths strive so hard that they join their peers in suicide pacts. But, don't be jaded. Do some introspection, look at yourself. Why do you wear the clothes you wear? Were the five year old clothes worn out or were they just out of style? Was your two year old car giving trouble or did you just see the "Joneses" pass in a new car?

Do you watch football on Sunday because

you really like it or because you like to share in the Monday morning camaraderie discussing those games? Does your son play high school football because he likes football, or is it parent pressure so the parents can be so proud of their athletic son? Does a young girl plan a wedding so soon after school because she is ready to spend the rest of her life with one young man or is it infatuation of a mother re-living her dream world? Does a Jewish boy have a Bar Mitzvah for his own good or is it the manifestion of the parents' desire to show off their young man to keep up with the "Jonesteins" who just had a half million dollar Bar Mitzvah for their son?

WHY DO WE DO THINGS? Is it inspiration from God? Is it because we feel good doing it? Or, do we act upon cues from the current trend? Consider this and think, "Would it be hard to get the majority of the world to accept the Mark of the Beast?"

Consider the many benefits of accepting the Mark. How much greater is the value of the Mark than the value of any fads of the past. If people don't accept it as a fad, would they accept it for security reasons?

In London I had my first hotel room with no key, at least no key of the traditional kind. My key was a piece of cardboard. On the inside of the cardboard was a coded magnetic strip. When I inserted the cardboard key into a slot beside the door, the door opened. When I checked out of the hotel the key was discarded. The next one using that room would be given a different piece of cardboard with a different code on the magnetic strip. The door receiver for the card key was reprogrammed for each hotel guest.

Later in Kansas City I was given another cardboard key to a hotel room. It had a series of holes. The holes allowed patterned light beams to pass. The light pattern created a code for the door reader. When the card was inserted into the door reader the door opened if the light pattern was correct for that room.

SECURITY! The major reason hotels use such devices is security. We may eventually want them for our homes. In many homes and businesses there are doors which open only by keying in a code at the door. These are used in many computer rooms.

I have been informed that one major U.S. corporation has a very secure computer room. A person must put his card key into a slot before the first door will open. When that door opens, the person enters a small chamber. Once in the chamber the door closes. If the person is the correct height and weight for the code on the card the other door will open and the person can enter the computer room. If the height and weight do not match the code on the card, the person is trapped and an alarm goes off. The person is prisoner until someone comes to open the security chamber.

TO THE NUDE BEACH

You can see security advantages of such things if you have ever gone swimming, jogging, or more extreme, to the nude beach.

Although the nude beach may be an extreme for some there are thousands of people that go to Black's Beach and others on warm days.

Imagine yourself on the beach naked. Where would you put your key and money for cokes?

With invisible tatoos, laser readers, biometrics, Point of Sale systems and security coded home and car doors, you don't need keys nor money!

Almost all shoppers and shopkeepers have a common, fearful question, "When will I be robbed?"

Without the Mark, "Anytime!"
With the Mark, "Never!"
How will the Mark avoid robberies?

With the invisible Mark and all other computer and communications technologies you can walk around stark naked without a purse

and still do any financial transaction from
buying groceries to renting a house. And, with
biometrics you can be verified or identified
for every transaction you make.

Pretend that you have been to the nude
beach. On your way home you would like to
stop at one of the drive in groceries like the
Farm Stores in Florida, but you have no money
with you. However, you do have the Mark. As
you drive up, the optical scanner reads the
Mark, 666, in your forehead. After the
computer verifies you as a member of the
Antichrist movement, the gate opens and you
drive through. The gate closes behind you.

You drive up to the window, extend your
hand and the reader reads the patterns in your
retinas. Then you make your order into a
microphone. The attendant presses the
necessary buttons and the total amount to be
paid is shown on a screen near your car
window. You say, "Okay." The attendant presses
another button and the computer mechanism
begins.

This Point of Sale mechanism read your
identification and sent the message to your
bank. When the money was transferred from your
bank account to the store account you opened
the little door, removed your groceries, and
drove home. The attendant never knew that you
were naked, but it didn't matter anyway. The
attendant was not human!

NOW, you have been initiated to the joys
and freedoms of the MARK! You may as well go
naked. The information bank knows everything
anyway! Even though the nude beach example
shows you a great carefree lifestyle, it also
portends the Beast. In this transaction the
date and time you were at the store were
recorded. You have a legal alibi. It is in
your records in the massive information
center.

Whether we see it as freedom or control
this is only the beginning of the Mark, but
don't forget, There are no free lunches. For
every benefit there may be a disadvantage.

But some in their religious fervor would say, "If you weren't wanting to do something wrong, you wouldn't worry about it." That is the exact logic that the Antichrist wants everyone to use!

Although true, it may seem to freedom lovers that all of our choices are now gone. Recently a law enforcement division paid $60,000 for a flashlight! Yes, a flashlight, but not the type you find in your local hardware store. This flashlight can spot you two miles away in the dark. That is not the half of it. It is infrared and creates no light visible to the human eye.

The benefits of this light are great. A person that just committed a heinous crime may be caught more easily. Or, it could lend a helping hand to those religious bigots who in a pretense of working for God would like to control everyone's life. For instance some of the electronic preacher giants who wanted to burn a few books or records could make use of the lights to spy on people in an attempt to monitor and control the lives of others through moralistic appeals.

In his autobiography Lee Iaccoca said that there was a rumor that Henry Ford II bought a listening device that could pick up voices through several walls. The device would allow him to eavesdrop on employees, especially those like Iaccoca. Is this Antichrist?

Are law enforcement officials, business giants, and lay church members bringing about the reign of the Antichrist? Are preachers themselves bringing about the rule of the Antichrist?

The answer is, "Yes."

Many of the preachers who are supposed to be heralding the cause of Christ, are the ushers of the Antichrist. Some leaders burn books that are on the shelves in school. It is a hate tactic, an effort to keep the young in the dark. If you're an adult and have not read <u>Catcher In The Rye</u>, pick it up and read it. Then tell me or yourself how old you were

Phonemes, Holograms, And Dirty Politics

before you knew all the dirty words that are used. Tell me how old you were before you heard these words. Tell me how old you were before you saw them scribbled on the bathroom walls. Is the book burning necessary? Is it effective? To effectively burn the book you would have to burn all school bathroom walls. Wouldn't you?

In at least one way the burning is effective? Many of these book burners get exactly what they want - publicity! It appears that the Bible is not worthy of their time. They need something else to serve their purposes. Some of these preachers could not get the attention any other way. Thank God there are so many of them seeking attention that soon there will be enough of the scenes that people will not even take a second glance.

Watch the scowl of these preachers when they perform. They don't preach that you go out into the world and put on a happy face. They seem to advocate fighting to force you to listen to their rhetoric about hell fire and damnation. Do they ever mention the love that Jesus taught?

Recently, on the steps of the nation's Capitol, a man was supporting a sign approximately five feet high and three feet wide. The sign had ten to twelve lines. These lines proclaimed the strength of Satan, the evil in the world, but only one line proclaimed Jesus as God. When I asked him why he didn't give God equal space he started to call me names. A Moslem friend with me asked, "Is he a Christian, a Preacher? Why does he do these things? Is it all for political reasons?"

These preachers preach that "When two are gathered in my name (Christ's name)" that all things can be accomplished. Yet, they are lobbying in Washington to stop abortion, to stop free choice of sexual styles, to reinstate prayer in schools, and running for political office. In California or South

Carolina you can watch the ministers on television yell about homosexuality as you sit and are sure that the guy preaching is the essence of the term "flaming queen". Then, lo and behold, you think the guy is bringing out one of the converts to testify. As you sit and wait for the testimony of the person who appears to be a two bit lady, he turns to her and introduces her as his wife. The next few minutes you run for the Pepto Bismol as she does a sickening performance of a testimony. You begin to wonder what this Christianity is all about. Then you ponder your own thoughts about the preacher and his wife. Is it the outward appearance?

The preacher tells you what a horrible condition this world is in. It is exactly as it was in the days of Lot in the twin cities of Sodom and Gomorrah and in the days of Noah.

Yes, preach on!

CHAPTER 14
THE SPIRIT OF THE ANTICHRIST IS HERE

Listen closely and you realize that the preacher read an episode in the Bible, but failed to get the message. Don't take the word of the preacher. See for yourself. Read the nineteenth chapter of Genesis. Is the story of Sodom and Gomorrah all about sex as your preacher says? Maybe he is so hung up on sex that it causes him to lose sight of the truth. Maybe he is so naive that he takes the words of a closet homosexual preacher. Does he think he will be vindicated of his own homosexual drives if he, before the world, decries such things? A friend once said of a famous (or infamous) televangelist, "I wonder about him. He never preaches a sermon without mentioning homosexuality."

Shakespeare said it aptly, "Thou dost protest too much." While growing up I heard it put another way, "The preacher is riding a hobby horse," indicating one of two things. Either the preacher has not studied for his sermon or he is guilty of what he is preaching. However, if the preacher is guilty or if he has not studied does not take away our own responsibility. We must "study to show" ourselves "approved."

Look at Sodom and Gomorrah. Just before the Beast years the world will be in the same state as it was then. C'est la vie! So it is! There is a preoccupation with sex, but that is not the problem. Now, as then, sexual promiscuity is merely a manifestation of the problem.

The sexual fantasy of the preacher, not the Bible nor God, seems to be the inspiration for his sermons. The basis of his sermons is similar to saying, "A person's sexual aberration for furry things is heterosexual because the furry thing that he cuddles is a doll of the opposite sex." This may seem to be an idiotic example, but think of that used to support the pulpit jingoists in their stories. In Genesis 19:5 the Sodom and Gomorrah story refers to the angels as men. The men breaking down Lot's door thought they were men, and today people refer to them as men. Yet, many paintings and wall hangings in churches and cathedrals portray angels as female. Think about this disparity and the veracity of your preacher. Is he obtaining inspired messages from God or is he merely trying to cover his own frailties?

The problem, of Sodom and Gomorrah and of today, is the lack of respect for God and fellowman. In the days of Lot no one respected the rights or property of others.

But, like Lot, the leaders of today are not very rational. They look for superficial answers and never get to the cause of the problem. Lot offered his virgin daughters to the men that were surrounding his house to rape the visiting "male" angels. Some ministers and leaders of today are just as ridiculous. They try to "cure" homosexual young men by providing female prostitutes. To stop the activities of abortion clinics some seem to advocate bombing the clinics, including the people. Is it, Support the bombers - murderers to get rid of the abortionist - murderers?"

The rationale does not seem Christian!

Technically, the mechanism for the Antichrist reign is here. More important though, THE SPIRIT OF THE ANTICHRIST IS HERE! The heralders of the Antichrist have their audiences right in the homes. Their unfounded teachings take the peoples' minds off the true teachings of Christ. Many of these preachers are antichrists. "...even now are there many antichrists; whereby we know

The Spirit Of The Antichrist Is Here 155

that it is the last time." (1 John 2:18)
Many times it seems that the truest of Christians are some of those of a faith other than Christianity. They have a feeling for their fellow man. They are the ones that sacrifice themselves to help mankind. Comparing these Christian- like, non-Christians to many of the professing Christians causes many to say, "I want no part of Christianity!"
Often the roles played and the sermons preached are totally devoid of anything Christ like!
Is your favorite preacher concerned with mankind when he collects millions in the name of God and sails out on his fifty foot yacht while many of those sending him their pennies are half starving? Is he serious in his critism of other preachers when he indicates his own humility and sacrifices saying that any preacher should be able to make it on his salary of only $100,000. But, forget the big time evangelist. What about your own pastor? Does he mention charity more often than he does tithes and offerings?
I hope your answer is, "Yes!"
The spirit of the Antichrist is here, not so much in the sexuality of the masses as in the pulpits of our churches.
Years ago many fundamental believers knew that the Beast mechanism would need fantastic record keeping. They also knew that this type invasion of privacy would be rejected by most of the population. However, for comfort and convenience people can rationalize, even to forgetting the negative points such as invasion of privacy. The Beast mechanism offers such creature comforts that it is being accepted without qualms.
The technology that controls people will improve their life qualities. For this reason people will not be forced to accept, but will ask the government for the Beast mechanism.
At first, the development of this technology was deliberate and slow. But, in

the last few years it has been so rapid that
nobody can keep up. As I write, there are so
many new advances announced each day that I
update this text regularly. It is intriguing,
fascinating, and frightening. It is an awesome
time for man. It is natural for the human mind
to wander and to wonder.

Having heard the Antichrist theory all of
my life, I look at every event with a wary
eye. With a somewhat scientific mind, I am
very curious. I do not accept everything at
face value. This skepticism and cynicism
spilled into my religious upbringing. This is
probably why I took note of developing
technology and saw it as a basis or support
for a full control government like the
Antichrist.

As you read the following descriptions,
place each event of this development within
the years of your own life. Consider how
impossible the Antichrist mechanism would have
been just a few years ago. Ask, "Is it now
reality?" Your own belief may not be altered,
but you may understand how others believe such
a theory. For the young, it may be difficult
to imagine life without these things.

My first experience with computers was in
a college engineering class. Although I hardly
knew what the word program meant, I was told
to write one. Within the semester I had
written a program in a language called
FORTRAN. The program solved a Thermodynamics
problem. When I wrote that program I had
really accomplished something. I did not know
what programming was. I did not know what
Thermodynamics was. And I really did not know
what FORTRAN was. Well, believe it or not, my
program worked.

What is a program?

A program is a set of instructions that
the programmer (a person) gives to the
computer. The totally stupid computer must be
told everything in extreme detail. If you want
a person to write his name on a piece of paper
you simply say, "Sign here."

The Spirit Of The Antichrist Is Here 157

To get a computer to do the same thing you must go through a lot of details such as: take pencil, put point to paper, move pencil, and on and on with details.
A computer language like FORTRAN is a set of instructions which the computer understands. The programmer learns the language to be able to give the computer orders. It is not necessary that the programmer really understand the total functionality of his program. This was true with my Thermodynamics program. I did not know the theory behind the problem or the equation, but I knew the symbols for divide, add, and other mathematical operations.
Daily it becomes easier and easier to tell the computer what to do. Thus, it is becoming very common for the person who barely knows what a computer is to use one effectively.
After my college experience with computers I got a job with a large construction company. I was a liaison between the accounting and computer departments.
To understand the tasks within my job, look around the kitchen, garage, office, or look at your holiday season card list. Just as you would list items around your home or office, I listed and organized invoice information.
This had been done using journals, ledgers and file cabinets. After the conversion, information was fed into the computer and it did the tedious additions and subtractions. It added columns and columns of figures in just a few seconds.
Everything was coded onto small forms. Keypunch operators then punched the codes into cards which the computer read. Once the cards were punched and read by the computer, reports were printed or stored on magnetic tapes. This magnetic tape is not much different from tape used in gift wrapping. It is magnetically treated instead of glue treated.
This system of coding forms, cards and magnetic tapes was an advancement over the

manual system, but primitive compared to systems of today.

My next exposure to computers was with an airline. The computer system was for the reservations offices. Here I saw improvements in customer service and record keeping. The real impact of the computer was vivid at the Christmas and Chanukkah rush season. This season provided a look into the future of the computer. One of my first customers confirmed my belief.

Within seconds after the customer gave me his name and flight number I confirmed to him that he, in fact, had reservations. He did not believe me until I told him his address and his telephone numbers. Then he exclaimed how great it was.

At his request I gave him a brief explanation of the computer system. He said that it had always taken thirty minutes or more to do what I had done in seconds. He was right. Many times prior to computers, agents never checked to reconfirm a persons reservations. It was physically impossible for the agents to file all of the reservations cards during the holiday seasons. The caller was simply told that he had the reservation.

My experience as a reservations agent gave me a basis for understanding the user's needs when I became a programmer. I saw the convenience and cost savings that the airline computer system would provide for the agents, the company and the customers alike.

Convenience and savings are important to everyone. If government and industry could pool their information without a public outcry, the Social Security number and your personal data could be taken from the census sheets and from your income tax returns and stored in a central data bank. A clerk or agent would no longer have to ask for names, telephone numbers, credit card numbers or other information. She would simply ask for your Social Security number. Would that not be very convenient?

Yes, it would be very convenient, but
dangerous. It is Beast perfect technology? It
will serve us well. It will protect us from
theft and other problems now inherent in some
convenient, money saving systems such as
credit cards.

This type of system fits into the Beast
structure. People will accept it without a
thought of the Beast.

You may ask, "Why has some company or
government not already made use of a central
information center?"

There are many answers:

1) It has taken more than twenty five
years to develop systems like the airline
system. The joint venture of the airlines
and a major computer manufacturer has been
modified for use with bank teller
machines, major credit and charge card
companies, train systems, hotel systems
and others. It appears that IRS will soon
use the same system.

2) There is bureaucratic pride among
companies. None want to lose control.

3) Government regulations often restrict
information sharing.

4) Some in corporate and government
management are stupid.

The list is endless, but problems disappear
daily as we see further development and
regulation changes.

It takes time! It took a long time to get
to the present state, but day by day things
are happening faster and faster.

To the non-believer, "The bureaucracy is
the main deterrent." But if God uses the
bureaucracy to govern the speed of
development, the answer may be the same as for

the believer: "God is not ready for it to happen!"

Watching the system develop over the years, I have seen continuous improvement. It has been intriguing and awesome. Companies have migrated from small computers to large ones. Not only has the size of the computers increased, the number of computers increased. Catalyzing this, technology increased capacity and power, and the Beast system drew nearer.

Today, the word processing computer that I used to write this book has more capacity than the first airline computers. It took massive buildings, special air conditioning, and humidity control for a computer system with no more capacity than my little table computer.

It is incredible and awesome! For the believer or non-believer we are on the brink of a Beastly system with or without omen!

For the non-believer there is no worry. For the believer there should be no worry. The fundamental interpretations of the Bible prophecies indicate that those who are 'living right' are okay! They will be taken away from the earth just before all of these horrible things take place. At the end of the seventieth prophetical week of years (7 Jewish calendar years), Christ will return to conquer the evil regime of the Antichrist. According to some, this will be the final blow to the Roman Empire and the beginning of the thousand year reign of Christ.

If a person truly believes this, it appears that he would take necessary precautions of prayer and study to ensure that he is 'living right.' If he is 'living right' the signs of the Antichrist should bring joy and peace because those signs are also the signs of the return of the Messiah, Jesus Christ. To many the description of Heaven as being pearl, gold, silver and other glittering things does not engender a desire to be in heaven forever. However, the thought of a thousand years reign and an eternal heaven where all is love, peace and harmony does

The Spirit Of The Antichrist Is Here   161

create an interest to many who face the hate, envy, and general lunacy of "modern civilization" each day.

Those who are truly ready have nothing to fear. Many have made comments about not wanting to deal with computers because of their BEASTLY nature. But, if it is prophesied by the men of God, it is going to happen. I certainly do not agree with one book I recently read. The author indicated that you should avoid everything that appears BEASTLY. According to him you should refuse to take a number at the laundromat or in a restaurant. It seems that God's people should never hesitate to use the tools of the Beast. They only need to be aware as they watch prophecy fulfilled. They should rejoice in the knowledge of the truth and be absolutely sure that they are "ready" when Christ returns to take the Saints away from the evils of the Beast.

If this sounds like preaching, it may be! I hate hypocrisy and to me the person that says, "I believe it," but lives as if he has never heard it is much worse than the one who says, "I don't believe it!"

Do you remember the Bible words "...better for the one who never...?"

Many airline managers have never heard of the Beast; yet, some fundamental believers may say that airline managers are intentionally working for the Beast.

This idea could probably not be farther from the truth!

Most managers use computers for one reason. That reason is profit! Computers help companies operate more efficiently and profitably. In some rare cases the top management may be on big ego trips, in which case the managers want large computers just to brag that their company is on the leading edge of technology.

Probably few of the decision makers even know the Antichrist philosophy. Could God allow people to fulfill prophecies while not aware of it?

One executive of a major bank in the United States had never heard of the Antichrist until I told him the subject matter of this book. Once I told him that my book deals with the Antichrist and the Mark of the Beast, I had to explain the philosophy.

The Antichrist society will develop smoothly. It will happen because of human ingenuity, curiosity, and a desire for a better quality of life. It will not develop because of evil thoughts of corporate managers!

Fire was not used the first time to harm anyone. The wheel was not made to defy God. Cars were not built to cause traffic deaths. Trains were not built to derail. Airplanes probably were not built to crash nor to provide faster transportation. For the Wright brothers at Kitty Hawk, North Carolina it was probably curiosity and challenge. This challenge was not even unique to the Wright brothers. It has been recorded in history for hundreds of years. For pilots Dick Rutan and Jeana Yeager the nine day Voyager flight around the world in 1986 was not an effort to defy God!

Many inventions have caused pain and suffering. Fire burns buildings and kills people. The wheel is necessary for most cars, trains, and airplanes which have all been a part of human disaster, but we continue to use them.

Some claim that the space program is an evil attempt to disprove the existence of God. In the case of most astronauts this is not even remotely true. Astronaut Frank Borman had personal devotions while the whole world listened as he floated around the moon. He even received criticism from some atheists for his prayers. Does he sound like a person trying to disprove the existence of God?

Some medical equipment, computer components, and many things we use in our daily lives, such as Tang, have been developed

The Spirit Of The Antichrist Is Here 163

as a result of our space program. Yet many still ask, "Why do we have a space program?"

Is it to defy God?

No! I think not!

We have a space program to satisfy the human nature of curiosity, ingenuity and challenge.
Many fundamentals may believe that we need to spend our time seeking the will of God. But, we may be fulfilling prophecies and doing the will of God. The airline system may even be a part of that will!

In that system, according to some estimates, there are only about 5,000 persons even remotely involved in its development and maintenance. But, many computing theories and ideas have been developed around or contemporaneously with the airline system. Many of the ideas are perfect for the Beast, but probably were arrived at without a thought of the Beast. Many of the other concepts are just as useful as the basic airline system.
One of the concepts is Timesharing. Although not a system per se, it is a concept in the efficient use of computers. Timesharing simply means that more than one person may share the computer resources. In the computer vernacular, each user has a slice of computer time. The main reason for the concept becoming popular was simple economics. Until the mid-seventies computers were so expensive that many companies and organizations could not afford them. However, through timesharing, the organizations could afford a "slice" of the computer as illustrated on the next page.

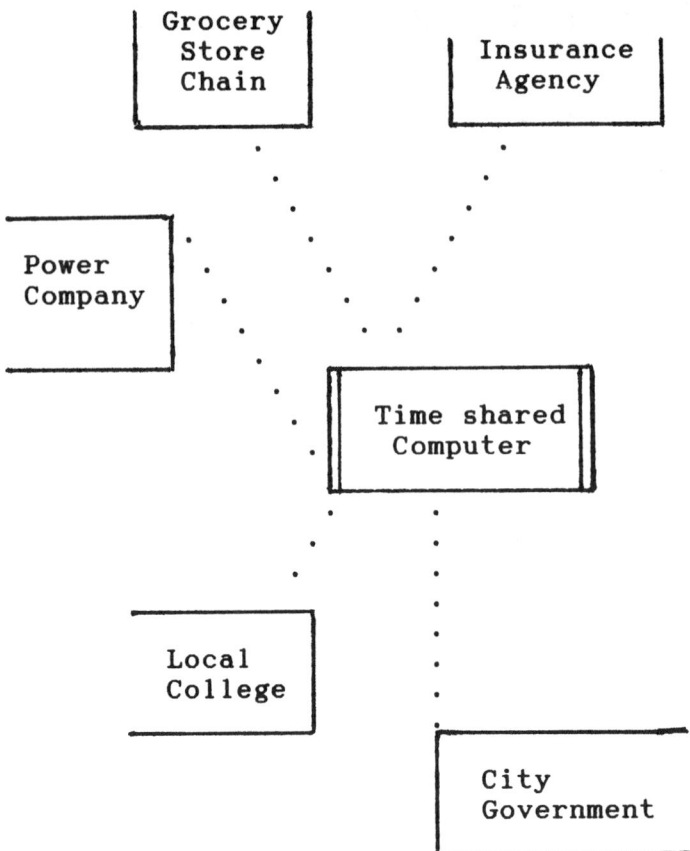

The Data Base concept also became popular. In fact, good systems with massive amounts of stored information must have some type of data base. When computers first began to be used, much of the data (words, numbers, etc.) were actually stored in the programs. Data bases were developed to avoid this. This provided two advantages. First, it made the programs more flexible and more useable. Second, it put the data in order in one central place. Having the data defined and stored in one central location made the data available to many programs.

An example of the improvements due to data

The Spirit Of The Antichrist Is Here   165

bases may be illustrated by the personnel data of a company. In the old approach, without data bases, the employee information was put into a table built within each program, thus creating a need for duplication of the data for every program that used the data. Using data bases the data is defined, labeled and stored one time. Each program that uses that data accesses the data from the same place rather than from within each program. In the early days of computers this was necessary to save valuable computer space. This concern diminished as computer storage space became cheaper and cheaper. However, good data bases have a more important reason for being. Data stored only once improves the accuracy and consistency of that data.

A very simple illustration may be made by considering a family of four. Pretend that cooking breakfast is a computer function. The stove is the computer, each person a program, and the cabinets and the refrigerator are combined to form the data base. The shortening, the sugar, the salt, the eggs, the bread, the meat and other food stuffs are the data.

Before data bases, each person of the household would have needed fully stocked cabinets and refrigerators. With the data base concept all of the food and all of the cooking utensils are in one area. This makes it easy for everyone to get to the needed supplies. It also takes less space. There is only one can of shortening for the group. The same is true in the data base. A person's name is stored only once, but all users may access it.

Another concept that incorporates many of the other concepts already discussed is the Information Center. The theory of the Information Center is similar, but somewhat broader than that of the Data Base.

The term Information Center simply means that the data bases have been expanded to include complete data of an organization. Technically speaking the word information means processed data. The Information Center definition seems to vary slightly for every person that has dealt with computer systems.

It may be meaningful to think of the Database and the Information Center as a three ring binder type notebook. You have done an inventory of the items in your kitchen. You have written this inventory list on one notebook sheet of paper. This sheet may be used as a check list for the member of the family that has been assigned the task of cleaning the kitchen. You have created a small database!

Now go through the house and create a Database for each room. In separate sections you put all of these Databases into the notebook. By each item on the inventory list you draw columns. You instruct the family member responsible for cleaning each room that the date of cleaning each item should be put into the column beside the item. You have now created an Information Center!

Your newly created information center with its collection of room databases seems simple enough. They appear to be of no interest to the Antichrist. However, assume that every household in the country did this. It would be a massive amount of information which could be of great importance.

It would amount to about eighty million households or notebooks. That is a lot of notebooks, but all of the information could be stored easily in many of the corporate or government computer systems. As simple as your three ring information center is, it could become an instrument of the Beast. Your information center combined with all others in the country could provide valuable information to the government and to the manufacturers of household goods. With this type information the government could allocate household goods. Eventually there would be a record of when you purchased everything in your house.

You can see the Beast more clearly!

It is easy to see that with all of the

The Spirit Of The Antichrist Is Here   167

computer, communications, and other high technologies combined that a Beast system can be implemented. For the Internal Revenue Service to catch all the cheaters it is necessary that all of your financial transactions be recorded. That means that each time you are paid it is recorded. Each time you purchase an item it is recorded. Each time you donate an item to the Salvation Army it is recorded. All of the transactions will be recorded under your Social Security number until biometric identification is implemented. As the population of the world grows and our natural resources become more valuable a system involving all technologies will be an absolute necessity.

One person asked me, "How will the IRS know when you donate a shirt to the Salvation Army? How will they know if you lied about its worth?"

It is simple once the intricate Beast technology is in place. Every organization and every home will have the biometric readers. Predictably that reader will be your television or CRT screen. Each time a donation of any item is made, it will be verified that you own it from the information stored about your previous purchases. If there is no record of your purchase and an expenditure you will not be allowed to claim it as a tax deduction. In fact, the information about your donating something that did not belong to you may be turned over to the law enforcement officers to investigate your activities in obtaining the item. And remember this may be done with NO HUMAN INTERVENTION.

From a security standpoint this is great. From a social standpoint it is definitely an improvement to our lives. And, from a technical standpoint it is exciting. But, from the standpoint of the fundamental believer it is BEASTLY! Everything in our lives will be monitored, from buying groceries to giving gifts. For the Antichrist mechanism to run swiftly and smoothly, one more thing must be done: The personal data on each of us world citizens must be gathered!

This process has started. It is working in many areas. Recently an announcement was made that government employees would have the convenience of automatic deposit of their 'paychecks.' In the envelope containing income tax refund checks there was a note to inform us that any dependents claimed who are age five or older must be identified by their Social Security Numbers. Even before this, it had started with our children. It was even advertised on large billboards!

The Spirit Of The Antichrist Is Here

CHAPTER 15
BILLBOARD
AND THE CHILDREN

Each day we read of children being abused, children being lost, children running away and in general the terrible plight of human beings from birth to death. While reading these gruesome articles we think, "What's the world coming to?" or, "We're in the last days!"

Although the thoughts vary, you never hear, "We need the Antichrist."

Yet, across the the land caring parents and concerned citizens are crying for the BEAST.

Immediately you say, "Oh, Yes, in San Francisco, New York, and other cities fraught with the plague of Sodom and Gomorrah, but not in the cities still occupied by the God loving, God fearing people."

How blind we sometimes are!

The cry is heard from within the quiet, sedate cities in the guts of the 'Bible Belt.' The ministers, the politicians, the parents and citizens of every walk of life cry out for the Antichrist.

Recently, in a Virginia town just outside our nation's Capital, I saw a shopping mall billboard crying for the BEAST. You could read it blocks away. Divided into two sections, one side advertised movies playing in the mall theaters. The other side was, in essence, a message from the BEAST.

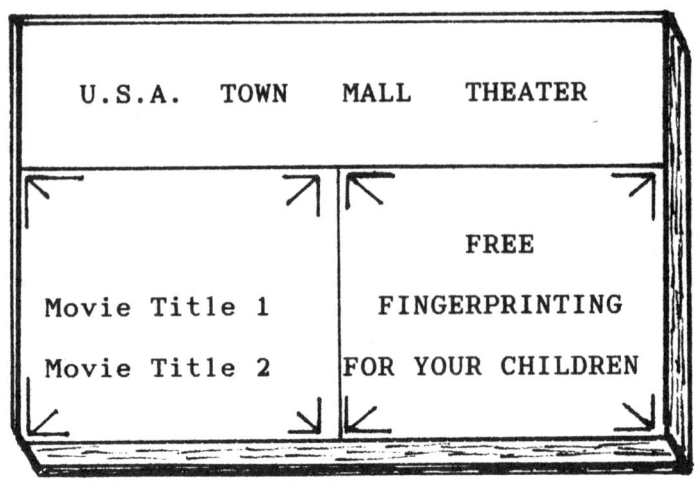

Yes, the sign was advertising fingerprinting for young children. Why is this happening? Why would anyone want to fingerprint their children? Are they criminals?
Absolutely not!
Local governments and citizens groups are trying to combat atrocities perpetrated against children. Fingerprinting is a way to track and identify children when they are lost, kidnapped or killed.
On August 27, 1986, a friend who helped edit this book asked if I had seen the Gaffney Ledger. It carried an article about fingerprinting children in Gaffney, S.C. She had also heard on the Spartanburg television station that South Carolina is considering Social Security numbers for all five year olds. Already in Holland, newborns are given an identification number. The Internal Revenue Service requires social security numbers for all dependents five years old or older claimed on your tax return filed in 1988 and later.
You may bristle and ask, "Is this so bad? Is this so BEASTLY?"

Billboard And The Children

Superficially, NO!
However, the BEAST is not the roaring, growling kind of a BEAST. He is a "Wolf in sheep's clothing." He will conquer people before they know his true identity.
We all may see the good in fingerprinting children. Most of us see some bad side effects. You may think of George Orwell's Big Brother (government) controlling our lives even to the point of sexual relations between a husband and wife.
It is hard to allow thoughts of Big Brother or the BEAST to overshadow thoughts of protection for our defenseless children. For the Christian there is a fundamental statement of Jesus Christ which demands that we seek to protect our children, "Suffer the little children...."
This is causing society to take a logical path to the Antichrist.
We see the wickedness of Sodom and Gomorrah sweeping the world. We review the last few years and witness how flagrantly churches, government, and the underworld are forming a mighty triad. As the wickedness develops, everyone must find ways to combat baby snatchers, burglars, and others violating our personal safety and security.
Simple acts like fingerprinting children fall into the logical progression toward the Anitchrist mechanism. Once older children are fingerprinted, babies will be identified as soon as they are born. This will eliminate the fear of hospitals giving babies to the wrong sets of parents. At first identification will be by parental consent, but soon afterwards it will be hospital procedure. It will be as automatic as circumcision for new born baby boys became after World War II.
Soon after fingerprinting at birth becomes customary, another sensible "at birth" happening will become automatic. The baby will be given a personal identification number or personal identifier. This personal identifier will be even more real and secure through the

use of biometric technology.
The BEAST progression continues.
The baby is forever identified.
The baby has a blood type, visible physical characteristics and even a family tree. It makes sense that this information be stored with the human identifier. With the technologies and concepts available the baby identification can be expanded.
As the baby develops and grows, information about the baby grows. The endless list of easily saved information will improve productivity in almost every area of society. There will be many cost benefits of the BEAST.

The following list is only the beginning of things that will be recorded.

<u>Individual Person
Information List</u>

Grades
Birthdate
Family tree
Demographics
Family Health List
Family Church Attendance
Family Legal Record
Marital Record
Friendships
Activities

It would take very little computer storage space to record these data for every person alive. It is a simple movement to protect our children, but a perfect step toward the Antichrist rule. In the age of paper record keeping it was impossible, but technology makes us aware that the Antichrist could begin his reign today.
Many books have been written about the Antichrist reign, but few, if any, have described the scope of technology that will be

Billboard And The Children 175

    the work horse or BEAST which provides the technical support for his reign.
    As several chapters of this book have left religious philosophy to dwell on high technology, The BEAST, it is easy to see how the Antichrist can keep up with you and control you. For years scientists have tagged everything from birds to caribou with sonar devices. This is done in an effort to track the animals to study them and their movements.

    In the November 1986 issue of "EQUUS" magazine there was an article, "Microchip Aims To Abolish Equine Identity Crises." The article describes how a veterinarian uses a 12 guage needle to inject an "ID Chip" two inches deep into a horse's neck directly below the hairline of the mane. According to the inventor, this chip will provide better identification of horses than lip tatoos or physical characteristics. Animals are tracked and identified.

Be sure that the Antichrist will track you,
and know everything about you.
You are identified and numbered —

666

It's all for comfort!
Can you take it in stride?

BOW DOWN BEFORE HIM,
STAND UP, ADORE HIM!

That is all that
is required
of
YOU!

REFERENCE AND READING LIST

Anonymous,.The "BEAST". circa 1979.

Anson, Jay. 666. New York: Pocket Books, c1981,1982.

Barnett, Donald L. and Njama, Karari. Mau Mau from Within. New York: Monthly Review Press, 1966.

Bell, Daniel. ed. The Radical Right: The new American Right. Salem, NH: Ayer Company Publishers, 1963.

Book Of Mormon, The. Salt Lake City, UT: The Church of Jesus Christ of Latter-day Saints, 1974.

Cantelon, Willard. The Day The Dollar Dies. Plainfield, NJ: Logos International, 1973.

Cerf, Christopher and Navasky, Victor. The Experts Speak. New York: Pantheon, 1984.

"Church Sexist." U.S. News and World Report, 12/17/84.

Codification of Presidential Proclamations and Executive Orders. Washington, D.C.: U.S. Government Printing Office. 1961-1981.

Commerzbank. EEC IN Figures.Washington,D.C.: Dist. by: Embassy of The Federal Republic of Germany. 1982.

Crozier, Michel;Samuel P. Huntington;Joji Watanuki. Trilateral Commission. New York: New York University Press, 1975.

Cumbey, Constance. The Hidden Dangers of the Rainbow. Shreveport, LA: Huntington House, 1983.

Eichelberger, Clark Mell. United Nations: The First Fifteen Years. New York: Harcourt, Brace, Javanovich, c1960.

Eichelberger, Clark Mell. United Nations: The First Twenty Years. New York: Harcourt, Brace, Javanovich, c1965.

Eichelberger, Clark Mell. United Nations: The First Twenty Five Years. New York: Harcourt, Brace, Javanovich, c1970.

Fehrenbach, T. This Kind of Peace. New York: McKay, c1966.

"50 Ways to Hide a Personal Computer." Management Technology. May 1983.

Freiberger, Paul and Swaine, Michael. Fire In The Valley. Berkley, California: Osborne/McGraw Hill, 1984.

Galbraith, John Kenneth. Economics & The Public Purpose. Boston: Houghton Mifflin Company, 1973.

Harris, Alphus L. "Timeline To Antichrist." Shelby, N.C: Joe H.E. Skates Publishing Company, 1985.

Harris, Alphus L. "Timeline To Eternity" Shelby, N.C: Joe H.E. Skates Publishing Company, 1985.

Heilbroner, Robert L. The Making of Economic Society. 2nd ed. Englewood Cliffs, N.J.: Prentice-Hall, Inc. 1968.

Hogendorn, Jan S. Modern Economics, An Introduction. Cambridge, MA: Winthrop Publishers, Inc., 1975.

Hunt, Dave and McMahon, T.A. The Seduction Of Christianity,Discernment In The Last Days. Eugene,Oregon:Harvest House Publishers, 1985.

Reference And Reading List

Iaccoca, Lee with Novak, William. IACCOCA. New York: Bantam Books, 1984.

Karmein, Joe R., ed. "Research Compendium." Shelby, N.C: Joe H.E. Skates Publishing Company, 1985.

Karmein, Joe R., ed. "Anthro-Chrono Apercus Ad Infinitum." Shelby,N.C: Joe H. E. Skates Publishing Company, 1985.

Larkin, Clarence. Dispensational Truth. Philadelphia, PA: Rev. Clarence Larkin, Est. 1920.

Lindsay, Hal. There's A New World Coming. Eugene, Oregon: Harvest House, 1973.

McConnell, Campbell R. Economics. Ed. 6 New York: McGraw Hill. 1975.

Microchip Aims To Abolish Equine Identity Problems. Equus, Nov. 1986.

Nave's Compact Topical Bible (The). Grand Rapids, MI: Zondervan Publishing House, 1972.

New Catholic Encyclopedia (The), Ed. staff of Catholic University of America. New York: McGraw 1966,1967.

"Optical Circuits." Business Week. Aug., 1986.

Pinkley,Virgil with Scheer, James F. Eisenhower Declassified. Old Tappan, NJ:Fleming H. Revell Co. 1979.

"Pushing Computers To The Ultimate Speed Limit." Business Week. 7/28/86.

Random House Dictionary(The). Stein, Jess, Ed.in Chief.New York:Ballantine Books,1980.

Roberts, Benjamin C., Ed. Trilateral Commission. Montclair, NJ: Allanheld, Osmun, c1979.

Roosa, Robert. Trilateral Commision. New York: New York University Press, 1982.

Samuelson, Paul A. Economics. New York: McGraw-Hill, Inc., 1970.

Scribner's, The Dictionary of the Bible. Ed. Grant, Frederick and Rowley, H.H. New York: Scribner, 1963.

Shultz, George P. and Dam, Kenneth W. Economic Policy Beyond the Headlines. New York: W.W. Norton Company, Inc. 1977.

Sigrid, Arne. United Nations Primer. New York: Rinehart, c1948.

Sklar, Holly. Ed. Trilateralism. Boston: South End Press, c1980.

Spencer, Milton H. Contemporary Economics. 2nd ed. New York: Worth Publishers, Inc. circa 1974.

Spencer, Milton H. Contemporary Macroeconomics. New York: Worth Publishers, Inc. 1975.

Weekly Compilation of Presidential Documents. Washington, DC: U.S. Government Printing Office.

White, John Wesley. The Coming World Dictator. Minneapolis, MN: Bethany Fellowship, Inc., 1981.

INDEX

abacus 71
Abed-nego 14
aberration(s) 136,154
abortion(s,ist) 69,70,
 150,154
Abraham 85
abuse(d,rs) 69,171
A.C. Bhaktivedanti Swami
 Prabhupada 131
act(s) 26,92,145,137,173
actors 106
Adam 63,65,108
adoration 64,86
adult 48,57,149
advertis(ed,ing) 171-2
advocate 38,154
Agca 26
agent(s) 83,89-93,95-8,158
agnosticism 31
Agriculture 23,69
Air Radio Incorporated 92
airports 89
airplane 89,92,123,162
AIRLINE(s) 83-4,87,89-99,
 101,105,110,114,122,
 124,140,142,158-160-3
America(n,s) 46-7,50,59,
 81,138,141
American Express 84
analog 66-8,70-1
ancestor 51
Anchorage,Alaska ANC 90-5
angel(s) 137,154
animal(istic,s) 49,136,
 175
animat(ed,ion) 144
Anson,Jay 105
anthropolog(ical,y) 11,109
anthropomorphic 44
ANTICHRIST(s) 10,12,13-9,
 20,21-5,27-9,31,34-5,
 40-1,45-7,56,63-5,68-9,
 75-6,81-3,85,90,95-7,
 102,105-9,112,114-6
 119,126-8,131,133,135
 -7,142,144,147,153-6
 161-2,166-7,173-6

Apocalyptic 27
Arab 128
Arabic 38,117
ARINC 92,94-5
Army 27,28,167
arsenic 43
Artificial 75,141-2
Aryan 58

assassinate 132
Assemblies of God 82
Assembl(y,er) 99,100,117
astrology 74
astronauts 162
AT&T 78
atheis(m,t)'13,31
ATL(anta) 89,90,92-94
atom 104
atroci(ty,ties,ous) 26,57
 60,69,70,172
audio 59
Austria(n) 22,58
author(ity) 19,20,35,58,
 60,103,136,161
Babbage,Charles 72,76
baby(ies) 69,96,173,174
bacteria 77
Bakker,Jim 26
bald 43
BANK(ing,s) 84,108-9,114-
 6,118-9,121-5,128,138-9,
 142,147-8,158-9,162
Baptist 127
Bar Mitzvah 145
baseball 103
BEAST(LY) 9-12,14-5,22,
 25,27-9,31-6,38,40-1,45-
 8,52,60,64-5,68-71,75-
 78,81,83,96-8,101-9,112
 -3,115-6,119,122-5,127-
 8,138-140,143-5,147,153,
 155,159-163,166,171-5
 166,171-5
beauracracy 112,159
bedfellow 31
Belgium 22
Bell,Alexander Graham 143
Bible 48-9,52,85,72,128,
 137,149,153,160-1,171
Big Brother 71,75,112,123
 173
bigot(s) 130,148
BILLBOARD(s) 168,171
billion(aire,s) 39,78,85,
 122,138,141
Biochip(s) 45,74,77,85
Biometric(s) 65,68,70-1,
 75,83-6,109,115,142,146-
 7,167,174
bit(s) 68,115,143,150
BhagwanShreekajneesh 130
BLACK('s) 56,63,97,129,
 130,132,146
Blacksburg (S.C.) 125

blind 57,109,129,130,171
blood 65,84,85,132,174
Bolivar,Simon 51
bon jour 118
Book of Mormon 15-6
Born(Again) 32,46,69,173
Britain 22
British 112
BROADCAST 72-3,131
buenas dias 118
bugle-like 131
burglar 108,173
Byron,Augusta Ada 76
Byron,Lord 76
byte 84
Calculus 67,72
California 74,90,132,149
camaraderie 145
campaign(s) 32,46,48,50
campus 58,60
cancer 64,70
Capital 69,171
capitol 149
car(s) 29,84,107-8,
 115,125,144,146-7,162
card(board,s)36,42,74,84
 110,119,125,145-6,157-9
caribou 175
Carson,Johnny 141
Carter,Jimmy 32
cash 84,113,116,119,125
Catalyzing 160
Catcher In The Rye 148
Catholic 13,18-9,24-6,82
 127,129,136
cattle 49
Central Processing Unit
 (CPU) 77
Chamberlain,Richard 16
Chanukkah 158
Charisma(tic) 17,33,108
Charleston 94
Charlotte Observer 70
Chase Manhattan Bank 84
Chicken 46
CHILD(REN)14,16,18,29,48
 52,69,103,168,171-4
China 40
Chinese 117
Chip(s)74-5,77-8,84-5,
 103,141,175
Christ(ian,s,ity)10,13-4
 17,25,46,49,51,53-6,69
 76,85,107,113,135,137,
 149-150,154-5,160-1,173

181

Christmas 158
Chrysler 138
Church(es) 18,24-26,31,33-4,42,82,102,107-8,128,
132,135-6,138-9,142,149,
154-5,173-4
Churchill,Winston 52
circuit(s) 72,77
circumcision 52,173
City of Seven Hills 21
defense(less) 106,128,173
CLIMATE(s) 46,135,144
code(s,ed,ing) 21,42,64,
70,86,94,102,113,139-40,
145-6,157
Cole-Whitaker 26
College 73,135,156-7,164
column(ist,s) 136-7,138,
157,166
comfort(s) 32,69,71,155,
176
Commerc(e,ial) 117,118,119
commingle 49
communication(s) 10,46,76,
96,112,118,123-4,146,167
communistic 27
Company(ies) 70,72,77,85,
87,97,141,157-161,163-5
COMPUT(er,ers,ing,'s) 10,
28,38,43,45-6,56,66-8,70
-78,85-7,89-104,106,110-
6,119,121-4,130,139-144
146-7,156-168,174
concept(ion,s) 69,86,110,
118,163-5,174
Conciousness 131
congressmen 20
Constitution 25
CONTROL(lers,ling,s) 28-9,
32-4,63,69,81,113,117,124
147-8,155,159,160,173,175
CONVENIEN(CE,s,t) 81,108,
110,113-4,119,156,159

Corporat(e,ion,s) 74,77,
86-7138,146,162,166-7
COSMETICS 56,64
Council 31
cow 51
Credit 32,42,74,84,86-7,
96-7,113,115-6,119,158-9
crime(inal,s) 58,66,110,
115,128,148,172
CRT(s) 45,92-3
Cuban(s) 19,20,50
Cumbey,Constance 26,107
currenc(y,ies) 36,120
daddy 9,76
Daniel 14,16,27

Darius,King 14,15
Dartmouth College 73
Darwinian 117
Data(base,s) 42,70,77,86,
109-10,125,158,164-5,168
Dear,(William) 103
death(s) 17,43,53,162,171
debit 113,115,121,125
Debra's Auto Parts 125
decadence 40-1,138
Training) 133
DECENERATION 101
Demographics 174
Denmark 22
Denver(DEN) 88
dependents 85,113,168,172
Depression 46,95,116
detective 103
deterrent 128,159
device(s) 143,148
Devil 17,46,49,50,106
Diabetes 49,70
DIARY OF ANNE FRANK 55
DICTATOR 17,19,37,40,124
DIGIT(al,ized,S) 36,47,
66-8,70-1,103,106
Digital Equipment(DEC) 74
discrete 67-8
disk 45,74,99,111
Divine Light Mission 132
dollar(s) 38,102,114-5,
117-9,122,133,145
dominant 137

drive(s) 84,99,110,129
drive in 147
Dungeons And Dragons 103
Earth('s) 28,48,51,53,76,
78,85,132,160
Eastern(Airlines)24,26,84
E-Church 138-9
ECONOMIC(ally,S) 11,22-3,
40-1,44,108-9,114,135
EFT 114,122
Egbert,James Dallas,III
103
eggs 165
ego(centric) 138,161
Einstein, Albert 38,104
Eisenhower, Dwight D. 18
elderly 81
elections 32,48,57,69,82
Electric(al,ity) 19,58-60,
72-4,76,78,99
Electronic(s) 46,73,101,
114-5,122-4,134-142,148
E-minister 138-9
emotion(alism,s)108,130
employ(ees,ers,ment) 97,99
101,116,129,148,165,168

encod(ed,ing) 70,102
encryption 71
Engineering 45,49,156
England 22,40
English 99,100,117,128
Environment(alists) 136
Equine 175
Erhard, Werner 133
Esperanto 117
EST (Erhard Seminar
eternal 29,49,53,106-7,
127,162
European Common Market,
European Economic Commu-
nity(EEC),Europe of Ten
21-3,28,38,40-1

Eve 63,65,108
evil 15,52,82,127,161-2
evol(ve,d,ution) 11,63,
82,109,117,138
Executive 19-21,34,162
experi(ence,ments) 43,
60,67,112,119,156-8
expert(ise,s) 65,101,104
Ezekiel 27,53
FAD(s)56-7,105,108,145
fake 97,127
Falwell,Jerry 26,130
Fanatic(ism,s) 128
Farm Stores 147
fate 66,75
father 9,10,71,76,78,104
Fatima 51,129
FDR 46-7,82
Federal 42,109
FERV(ENT,LY,OR)108,129,
130,148
financial(ly)11,44,51,69
124-5,135,158,147,167
Finger(printing,s) 38,39
65,71,172-3
fire 17,35,138,149,162
First Interstate Bank 84
flaming queen 150
flashlight 148
flight(s) 88,90-1,93-4,
96,158,162
flirt 31
Florida 88,91,147
Flynt,Larry 170
Food15,19,20,101,108,165
FOOLPROOF 65-6,108
football 144-5
Ford,Gerald 132
Ford II,Henry 149
forehead(s) 21,29,38,56,
64,105,147
Foreign(er) 50,54,69,135

INDEX

forge(ry) 70
Framingham, MA 70
franc 119
France 22
FREE(dom,s) 17,22,99,147,
172
French 117-9
Fromme,Lynn (Squeaky) 132
Fulkerson,John 49
fundamental(ly) 10,13-4,
18-9,24,32,35,41,48-9,56,
76,81,83,107,113,127,129,
132-4,136,160-1,167,173
Futrell,Mary Hatwood 56
Gaffney Ledger 115,172
gay 58
Gun(s,tic) 38,45,49
Genesis 137-8,153-4
gentiles 55,127
German(y) 22,118-9
Gibson, Mel 16
Glasgow 118
Glove 57,60,105
goal(s) 23,31,38,138
God(liness,ly,'s,send)13-6
26-8,33,46-7,49-51,54-6,
64,107-9,112-3,127-9,
131-2,139,140,144-5,148-
9,154-5,159,161-3,171
God-son 51
Goy 27
Gold(en) 29,116,133,160
GOVERN(MENTS) 17,19-23,26-
8,35,38,41-2,44-6,52,54-
5,81-2,102,109-2,114,123,
132,135,155-6,158-
9,164,166,168,172-3
grafts 74
Graham, Billy 26
grain 49
graphics 144
greed 138
Greek 118
grocery(ies) 64,101-2,
113-4,119,121-2,147,
164,167
Gross National Product
(GNP) 23
guage 175
Guru (Maharaj Ji) 130,132
guton morgen 118
Guyana 130
HAIR(s) 35,42-4
hairline 175
HAND(SOME,s-on,writing)
29,41,57,64,112,127,
121,125,131,140,147
Hari Krishna 24
Harris Auto Parts 125,142
hat(ed,ing) 50,54,138,161

148,162
HEAD(aches) 105,117,125,
128,143
Heart(less) 52,66,70,75
heathen(s) 52
Heaven 17,51,160
Hebrew 14,16
hedonism 138
heinous 148
hell 149
Helms,Senator Jesse 69
homophilia 49

heterosexual 154
hexadecimal 39,40
Hindi 117
Hindu 13,136
Hitler,Adolf 57-8
Holland 22,172
Hollywood 118
HOLOCAM(S) 144
Holy 51
home(s)107-8,113,139,140-
2,143,146-7,155,158
Homer 32
homosexual(ity) 150,153-4
HORROR(s) 55,58,76
horse(s,'s) 153,175
hospital(s) 43,99,173
hotel 145-6,159
house(d,holds,ing,s) 77,
102,115,136,147,165
Houston 90
Hudson, Rock 16
Hula Hoop 50
HUMAN11,15,16,36,39,49,52
58,64,67,69,70,74-6,90,
95-6,102,114,116,140-1,
143,147-8,156,162-3,167,
171,174
Hungary 22
Hustler Magazine 130
hypocrisy 75,162
Iaccoca,Lee 138,148
IBM 72,74-5,103
ID(ENTIcal,FICATION,fied,
fier,fying,s,ty)65-5,68,
70-1,83-4,86,102,109,121
140,147,167,172-6
IGNORAN(CE,t) 9-11,33,40,
45,100-1,105,107-8,112
imposter 48-9,55,127
Inc.(Mr) 141
India(n) 117,131
industr(ialist,y) 75,87,
88,112,114,124,138,158
inerrant 49
inflation 103
inform(ation,ed) 40-3,45,
58,68,70,77-8,81,84-5,

87-8,91-3,98-9,101,123-
5,147,158-9,167,174
Information Center 66-7,
109-11,123-5,164-7
infrared 148
inject 175
ink 63-4
instrument 166

INTEL 73-4,78
Intelligence75,106,141-2
International 36,38,57,
83,117-9,122,144
Internal 109,167,172
INTERVEN(ed,TION) 95-6,
129,140,167
INTRIGU(E,ing)10-1,65,68
93,109,112,115,156,160
Invention 142-3,162,175
inventor(y) 88,93-4,166
Iranian 128
Ireland 22, Irish 18
IRS 168
Israel 17,28,54
Italy 22
Ivan the Terrible 17
Jackson, Michael 105
Japanese 118,120,138,141
Jesus (Christ) 13,25,43,
49,50,54-5,126,149,160,
173
Jew(ish,s) 16,18,53-6,
58,127,136,145,160
jewels 29
jingoists 154
Jobs,Stephen 73
jog(ging) 131,146
Johnny Carson,Inc. 141
Jonus,Jim 130-1
Joneses 144
Jonestein 145
Journal(s) 139,158
Judah 17
judgement 102
jungle 11
Justice 109,117
Kalispol,Montana 125
Kansas City (MAC) 88,145
Kennedy,John 19,34
key(board,ing,punch) 92,
106,139,145-6
kidnapped 172
kill(ed,s)60,130,162,172
Kindergarten 136
King Darius 14
KINGDOM 12,21,28
kitchen 157,166
Kitty Hawk,North
Carolina 163
kopecks 114

INDEX

Koran 128
Korea 132
Krishna 132
Kruschev,Nikita 40
Ku Klux Klan 42
Laboratories 49,78
Lady 76,150
Landers,Ann 136
language(s) 73,99-100,
 117-9,129,144,156
laundromat 161
laser 64,68,146
law(makers,yers) 48-9,54,
 136,148-9
leader(ing,s,ship) 34,45,
 52,58,69,108,118,130-3,
 148,154,162
League 28,105
Legal 69,147,174
legislation 69
liberty 50
library(ies) 78,140-1,143
license 84
life(style) 66,69,71,81,
 83,105,128,137,145,147,
 155-6,162,171
light(ning,s) 51,53,64,78,
 106,131,145-6,148
lion's den 14
lip 175
LISA 74
List 174
Little Rock 125
logic(al) 64,72-3,173
London,England(LON) 98,145
Lord 16,132-3
Los Angeles (LAX) 89,
 90,92-3
Lot('s) 136-7,150,154
lov(e,ers,ing)137,160,171
Lucifer 25
ludicrous 10-1
Luke 25,32
lukewarm 128
Lung 70
Lusaka,Zambia 123
Lutheran 26
Luxembourg 22
Machine(s) 72-4,76-8,
 124,159,167
magnetic 84,99,145
Magog 27
Maharishi Mahesh Yogi 132
Mail(box,ed,ing) 95-7,
 109,123-4,139-42
manage(ment,rs) 82,100,
 159,161-2
manifestation 136,145,153
mankind 47,155
Manson, Charles 18,132-3

Mariel 50
Mark (of the Beast)12,15,
 17,21,28-9,36,45,47,52-3
 56,61,69,70-1,81,90,102,
 105-6,108-9,112,118-9,
 127-8,145-7,162
Maryland 92
mascot 27
masturbating 57
mathematics(al)67,117,157
Mau Mau 52
Matthew 32,76
mechanism(s) 61,63-4,71,
 75-6,81,83,90,97,108-9,
 112,114-5,124,128,142-3
 147,154-6,167,173
medical 29,52,162
Meditation 132
Mediterranean Sea 22
Meir, Golda 18,127
Mengele 66
mental(ly)58,85,97,72,137
Meshach 14
Messiah(s) 13,17,47,54-5
 75,127,132-3,135,160
Miami,Fla.(MIA) 89-93,98
Microchip 175
microcomputer 74
microphone 147
microprocessor(s) 75,103
Middle East 22,127
military 109
million(aires,s)68,82,99
 117,139,141,144-5,166
minist(er,s,ry) 129,140,
 150,154,171
minoxidil 43-4
mirrors 78
mockery 131
modems 123-4
Mohammed 128
molesters 69
MONE(Y,y,less,tary) 77,84
 108,113-6,118-122,124-5,
 128,130,138-9,146-7,159
moon 48,162
Moonies 24
Moon,Sun Myung 18
Mormon 16,24
Morse,Samuel 72
Moslems 13,128-9,135,149
Mother 49,51,75,145
motto 138
murder(ers,ous) 60,132-3,
 135,154
mythical 106
naked 63,146-8
nanoseconds 125
narrow-minded 10
Nation('s,wide)28,130-1,

 141,149,171
National31,54,56,122,135
NATO 28,37
Navasky,Victor 104
Nazi 57,60
Nubuchadnezzar 16
Netherlands 22
NETWORK(s) 25,90,123,
 138,140,143
New Age Movement 25
newborns 172
New Jersey 89
newspaper(s) 46,112,136
New York (City) 40,90,98
 110,123,125,128,132,171
Nigra,Thomas 43
Nineteen Eighty Four 112
Noah 136,150
North Africa 22,27
North America 98
North Carolina 69
notebook(s) 166
nuclear 104
nude 146-7
NUMBER(ED,ing,s)43,47,52
 64-5,77,81-8,91,96-7,102
 105-6,109,117-8,121,129
 132,142,158,160-1,168,
 172-3,176
numerals 38,56
nuns 132-3
Occult 17,31
OCR 101
Offic(e,ial,s, 19,90,95,
 117,124,135,148,150,157
Office Automation(OA)123
ohayogozaimasu 118
oil 127
omen 102,160
ominous 13,14,122,123
Optical 75,101-2,119,
 125,147
optometrist 104
Orders 20-1,109,147
ORGANIZ(ATIONS,ed) 65,99,
 102,110-2,128,133,165
ORWELL,George 71,112,173
Pac-Man 103,105
pain 162
paper(less)50,114,157
paradise 47
parent(al,s)69,104,145,
 171,173
parody 130
parole 133
pastor 155
pay(ing,ments,roll) 82,
 97,121,102,116
PC(jr) 74,103
peace 16-731,36-7,69,109

INDEX

peacock 131
Pearl Harbor 41
pearl(y) 127,161
Pentacostals 127,129
PEOPLE 11,13,69,54,71,77-
8,84-6,115,130-1,135-8,
144,149,154-5,161-2
Peoples Temple 130
Pepsi Cola 114
Pepto Bismol 150

Perdition 18
philosoph(ers,ical,ies,y)
10-1,31,68,106-7,128,136,
161-2,175
Phobia 50
phone(s) 78,117,121
PHONEME(S) 143-4
pimp 33
Pilots 117,162
plague 171
pleasure(s)116,124,137-8
PL/I 73
Point of Sale 113-4,118-9,
121,123,125,140,142,146-7
police 65,109
POLITIC(al,ally,ian,S) 11,
13,22,27-8,35,49,52,54-5,
57,82,109,117,124,129,
135,144,149,171
Pope (John Paul II) 24
Portugal 51,129
POWER(S) 31-2,34,55,124,
126,139,160,164
pray(er,s,ing) 51,129,140,
149,160,162
preach(ed,er,ing,s)10,129
131,133,135,138-141,143,
148-9,150,153-5,161
predecessor 52,65
predicate calculus 72
President(ial) 18-20,25,32
34,46,48,56,57,69,82,132
142
Prince 18
Princeton University 48
prisoner 146
program(med,mer,ming,s)
46,52,73,76-7,82,99,100,
103,105-7,119,139,140,
150-8,162-5

progress(ion)72,76,173-4
Prolife 136
promiscuity 153
prostitute(s) 31,33,154
Protestant 135
Psycholog(ical,ists,y)
103,133
Public(ity) 20,21,34,136

149,154
Pulitzer Prize 50
pulpit(s) 107,154-5
pun 69
punishment 57,69
puzzle 46,56,114
QUALIT(ies,Y) 69-71,81,
83,105,128,156,162
quotas 55
Rabbi 129
rac(e,s,ial) 47,50,114
radio 72,76,107,139
Radio Shack 74
rape 154
rastafarian 132
rays 64
read(er,ing,s,y)64,71,117
119,135,140-1,146-7,
161,167,171
Reagan,Ronald 32
rebel 129
recipe 110
record(s,ed,ing) 68,70,72
81-2,108-110,143,147-8
155,158,162,166-7,174
recreation 29,103
recruiter 129
Redford,Robert 16
re-entry 63-4
refrigerator 165
regime 57,160
reign 15,31,47,56,63-5,76,
127,135,148,154,160,174
reject(ed,ing,ion) 14-5,
53-5
relative(s) 52,109
RELIGI(ous,OUS)10-1,22,24
-6,33,51-2,55,69,100,109
128-9,135,148,156,175
research(ed) 75,78,141
reservation(s)88,90-1,95-
7,105,114,158
restaurant(s) 84,161
ruines 64,71,147
Revenue 109,167,172
revolution 101
Reynolds,(R.J.) Inc. 70
Richmond,VA (RIC) 90,92-3
Right(s) 68-9,76,108,119,
137,154,158,160
ritual(s) 51,55
robbed 140
Rockleigh,N.J.(RNJ) 89
Rogers,Buck 48
Roman(s)18,19,22-6,28,40-1
82,127,136,160
romance 23
Rome 21,23
ROOM(s)58,63,77,99,145,166
Roosevelt,Franklin Delano

(FDR) 46-7,52
Rosenberg, John Paul 133
troubles 114
Rulu(s)45,51,106;133,175
Russia(n) 27,40,114,117
Rutan,Dick 162
Ryan, Leo J. 130
safety 115,173
Saints 76,161
Salt 114
Salt Lake City 90
Salvation 46,82,129,167
San Diego 131
San Francisco 171
Satan(ic) 13,17-8,150
satellite(s) 98,124
satrap 14
savior 47
School(s) 57,135,145
Scien(ce,tific,tist,s)10
45,46,49,50-1,59,65,68,
135,144,175
Scotland 118
Scribner's,TheDictionary
of the Bible 13
SDR 119
Sears 42,84
SEA(ttle) 90,93-4
sect 132
SECURE(ITY) 26,32,
81-7,101,108-
114-6,145-6,158,
168,173
Senate 70
sermon 116,130-1,139,140
sex(ual,uality) 136-8,
153-4,155,173
Shadrach 14
Shakespeare 153
Shelby,North Carolina130
skulls 114-5
shelter 108
shortening 165
shrine 51,129

SIGN(S)47,50,104,156,160
signature(s) 65,68,70
SILICON 77-8,85,103
Sin(s) 18,28,138
sinews 65
slice 103
Smith,Joseph 16
snatchers 173
SOCIAL (Security)46-7,52
55,69-72,81-7,101,109
111,135,143,144,158,167
172
society 50,60,70-1,101,
114-6,135,137-8,
162,173-4

sociology(ists) 45,104
Sodom and Gomorrah 136-8,
150,153-4,171,173
software (programs) 77,
100,104,144
sonar 175
Southern(ers) 54,127
South Pacific 140-1
Spain 22
Spanish 117-8
Spartanburg SC 172
SPIRIT 133,154,169
Spiritual 107
Springdale,Ark.125
spy 148
Sri Lanka 71
stagecoaches 116
Star Wars 44,48,75
Station(s) 89,172
St.Elsewhere 58
St.John 127,129
stor(age,ed,ius,ing,front)
  9,60,68,70,78,84,99,100,
  103,115,121,123,148,157-8
  164-7,174
stroke 70
suave 16,23,139,144
sugar 77
suicide 103,131,145
Sun Myung Moon 18,132
sun 48,140
super-human 13
SUPPORT(ers) 76,132,139,
  156,175
Supreme 76
surveillance 101
symbol(s) 47,99
synthesis(ed,ing) 144
System(s) 43,52,63-5,74,77
  -8,85-7,88,97,99,101-3,
  108,110,114,118-9,122-5,
  140,158-160,163,167
tagged 175
Talmud 103
Tang 162
tape(s) 99,157
Tate 132
tatoo(s) 21,64,146,175
tax(ed,es) 102,107,132,
  150,167,172
teachings 49,76,137,154
technology(ies,ical) 10-
  1,37-8,45-6,56,64-6,68-
  73,75-8,83,85,93,98-9,
  101,109,112-3,115,122,

126,135,141-2,144,146,155
  159,160,162,167,174-5
Teenagers 144
telegraph 72
telephone(s)54,72,77,89,97
  110,121,123-5,142-3,158-9
television 50,54,91-2,
  106-7,118,131,138-9
  144,150,167,172
teracts 78
Terminal 87,91,92,123,125
Texas 103
Texas instruments 75,141
THEATER(s) 171-2
The "BEAST" 48
THE DUNGEON MASTERS 103

Thomas,'doubting' 11
tickets 96-7
TIME(LINE,S,piece) 47,50,
  54-6,72-4,78,85,93,96,97
  103,128,141,147,149,
  155-6,159,162-3
Timesharing 163
TM 132
tobacco 70
toes 85
TOKA TOKA TOKA 41
Toronto 131
TOWN 171-2
track(ed,s)9,108,172,175-6
train(s,ing) 100,116,123,
  159,162
Transcendental 132
translator 73,77
transition 85
Transportation 136,162
transference 50
trend 107,145
Tribulation 16
Trilateral Commission 28
triticale 49
TRS-80 74
TRW 42,86
Tulsa (TUL) 89,92
Turkey 22
tyranny 13
ugly 133
underworld 173
Unification 132
United Nations 28,58,117
United States (of America,
  U.S.,U.S.A.) 25,32,40-2,
  45-6,50,54,59,77,83,90,96
  103,111-2,114-7,123,162

128,138,140,146,172

UNIVERSAL 83,102,113,
  117,119
Valentino,Rudolf 16

Vatican 25-6
veterinarian 175
victim 59,60
video 92,139,140,143
Viereck,Peter 50
virgin 154
Virginia 171
volt(s,age) 59,60
Voyager 163
Waldheim, Kurt 58
Wall Street Journal 138
Wang,T.C.138
war(s) 57-8,102
Washington, D.C.91,150
Washington Post(Mag.)
  43,49,56,78,130
webbing 65,71
West Germany 22
wheel 162
White 57,60,132
whore (Scarlet) 12,23-8
  31-34,41,44
wine 51,55
wizard(s) 51,135,139
Wolf 173
woman 51,136

WORLD(s,'s,wide)10-1,13
  22,33-4,37-8,40-1,44-5
  51-2,55,57-8,60,77,83
  86,98,100-9,112,114,
  117-19,123-4,133,135,
  138,145,153,162,167,
  171,173
World Common Market 22
World War II 41,58,173
Wozniak,Stephen 74
Wright brothers 162
xeno(phobia) 50-1
yacht 155
Yahweh 14
yak-tail 131
Yeager,Jeana 162
yen 119
Zambia 123
4004,6000,8000 73,77
666 9,21,29,47,64,71,
  86,105-6,115,147,176

ABOUT THE AUTHOR

Alphus Harris has twenty years experience in the computer business. He has been involved in banking, trains, airlines, hotels, construction, auto parts, education, and real estate.
His work endeavors have involved: marketing, marketing research, consulting services, retail, wholesale, programming and management.
With a zeal for truth, and a burning curiosity, he is continuously studying and travelling. His travels include more than forty of the United States and as many countries including a peek behind the Iron Curtain, walks along Copa Cabana and Waikiki, boating in the Greek Islands, pondering the Tower of London, meditating in the Sistine Chapel and Shrine of Fatima, admiring in El Prado and the Metropolitan Museum of Art, climbing the steps of the Statue of Liberty and the Washington Monument, and traversing the Way of the Cross.
He has earned the degree of Master of Business Administration and a Bachelor of Arts in Industrial Psychology. Degree, special studies and graduate work have included: Florida International University, New York Institute of Technology, Harvard University, Catholic University, Clemson University, University of Texas, Berry College, Central Piedmont Community, Dade Junior College, Greenville TEC, Airline Schools, training courses of IBM, Unisys, and DEC. These studies have included computers, communications, languages, engineering, religion and even lobbying.
Additionally, he has taught in several universities and colleges. The courses taught were: business, technical writing, computers, and even English to foreign students.

About The Author

While spending most of his lifetime in South Carolina, North Carolina, Georgia, New York Florida, Texas, Washington, D.C. and California he has experienced many of the different U.S. subcultures.

Spending his formative years in the "Bible Belt," provided him with a basis for comparing his studies, his experiences, and the prophesied scenarios for the 'End Times.' The experience, education, and travel provided a basis for comparing the reality of technology with seeming vagaries of Biblical prophecy.

He enjoys linguistics, swimming, biking and, of course, writing. But, a favored pastime is observing people, their religious attitudes, and the world in general.

Other Books By Alphus L. Harris:

Concepts And Computers - A collection of brief descriptions of computers and the concepts being used or developed today. Easy to read, it is a book for the layman and for the computer person who has not been introduced to all of the computer and related concepts.

Timeline to Antichrist - A news chronology for the person who does not have time to do the research for his next speech or sermon. Each entry is simple to understand.

English for Manager And Minister - A series of simple illustrtions to help the person writing or speaking to avoid appearing ignorant. It is powerful and simple. Almost everyone will improve before his next speech just by quickly scaning this book.

Spanish:Beginning Help - A short collection of hints for understanding Spanish. It is a good supplement to the beginning Spanish course.

Timeline to Eternity -A chronological research collection of news events, some of which are annotated for speech and report preparation.

Technology Gospel - A short,easy to understand collection of succinct descriptions of various technical developments that affect our lives.

REBB: Real Estate: The Boom,The Bust - This is a light hearted approach to the fantasy that one cannot lose in Real Estate.It explores the ways Real Estate agents may profit from real estate while no one else does. Finances, psychology, and money making myths are all discussed. A must for anyone hoping to go into real estate for home or for profit.

Also from Joe H.E. Skates Publishing:

Christians On The Seven C's by: Myra D. Goforth - Seven powerful lessons for promoting Christian Maturity. Each word in this book has a specific purpose. It is interesting and very rewarding when the lessons are followed daily.

Anthro-Chrono Apercus by: Joe R. Karmein - A compilation of brief descriptions of events of science, religion, politics and anthropology. Excellent for ministers, report writers and speech makers.

If you have any comments or questions about
the book Antichrist, Computers, And You!
write to the author at:

>
> Alphus L. Harris
> c/o Joe H.E. Skates Publishing
> P.O. Box 848 - A
> Shelby, NC 28150

Check with your local book stores. If they do
not have our books, please complete the form -
you may make copies of the form - on the next
page and mail to us at:

>
> Joe H. E. Skates Publishing
> P.O. Box 848 - F
> Shelby, NC 28150

Please allow four to six weeks for delivery.

ORDER FORM

Please send the following books by Alphus L. Harris:

_____ copies of Antichrist, Computers, And You! @$ 6.95 each.

_____ copies of Concepts And Computers @$19.95 each.

_____ copies of Timeline to Antichrist @$ 9.95 each.

_____ copies of English for Manager And Minister @$ 5.95 each.

_____ copies of Spanish:Beginning Help @$ 5.95 each.

_____ copies of Timeline to Eternity @$ 9.95 each.

_____ copies of Technology Gospel @$ 9.95 each.

_____ copies of REBB: Real Estate: The Boom, The Bust @$ 9.95 each.

Please send the following books by Joe R. Karmein:

_____ copies of Anthro-Chrono Apercus @$12.95 each.

Please send the following books by Myra D. Goforth:

_____ copies of Christians On The Seven C's @ 4.95 each.

Total amount enclosed: $ _____ . ___

Name: _____

Address: _____

City: _____

State: _____ Zip: _____

Shipping: $1.00 for the first book. $.75 for each additional book. S.C. residents add 5% tax. If you prefer faster service send $2.50 per book for Air Mail.